HSC Year 12
PHYSICS

PAUL LOOYEN | SANDRA WOODWARD

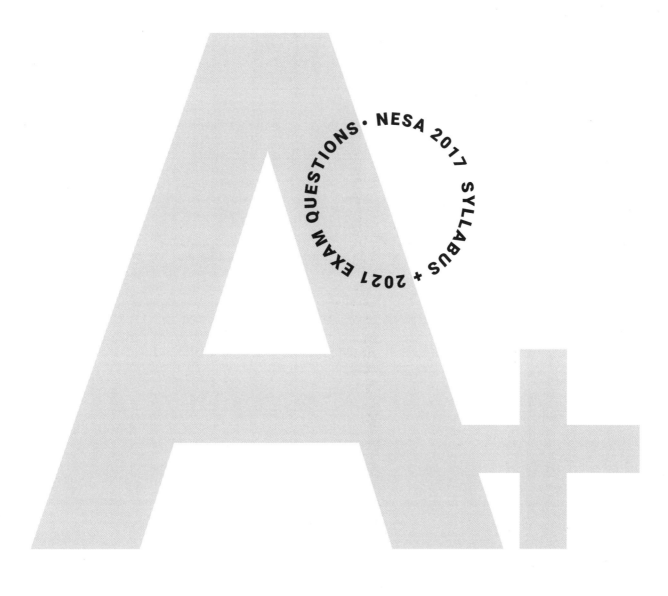

NESA 2017 SYLLABUS + 2021 EXAM QUESTIONS

A+

+ 16 topic tests
+ two complete practice exams
+ detailed sample answers

**PRACTICE
EXAMS**

Nelson

A+ HSC Physics Practice Exams
1st Edition
Paul Looyen
Sandra Woodward
ISBN 9780170465298

Publisher: Catriona McKenzie
Series editor: Catherine Greenwood
Copyeditor: Sam Trafford
Reviewer: Martin Barkl
Series text design: Nikita Bansal
Series cover design: Nikita Bansal
Series designer: Cengage Creative Studio
Artwork: MPS Limited
Production controller: Karen Young
Typeset by: Nikki M Group Pty Ltd

Any URLs contained in this publication were checked for currency during the production process. Note, however, that the publisher cannot vouch for the ongoing currency of URLs.

© 2022 Cengage Learning Australia Pty Limited

For product information and technology assistance,
in Australia call **1300 790 853**;
in New Zealand call **0800 449 725**

For permission to use material from this text or product, please email
aust.permissions@cengage.com

ISBN 978 0 17 046529 8

Cengage Learning Australia
Level 7, 80 Dorcas Street
South Melbourne, Victoria Australia 3205

Cengage Learning New Zealand
Unit 4B Rosedale Office Park
331 Rosedale Road, Albany, North Shore 0632, NZ

For learning solutions, visit **cengage.com.au**

Printed in China by 1010 Printing International Limited.
1 2 3 4 5 6 7 26 25 24 23 22

CONTENTS

CHAPTER 1

Module 5: Advanced mechanics

CHAPTER 2

Module 6: Electromagnetism

CHAPTER 3

Module 7: The nature of light

CHAPTER 4

Module 8: From the Universe to the atom

Practice exams

HOW TO USE THIS BOOK

The *A+ HSC Physics* resources are designed to be used year-round to prepare you for your HSC Physics exam. *A+ HSC Physics Practice Exams* includes 16 topic tests and two practice exams, plus detailed solutions for all questions. This section gives you a brief overview of the features included in this resource.

Topic tests

Each topic test addresses one inquiry question of Modules 5–8 of the syllabus. The tests follow the same sequence as the syllabus, starting with the first inquiry question of Module 5 and ending with the final inquiry question of Module 8. Each topic test includes multiple-choice and short-answer questions.

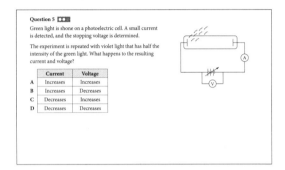
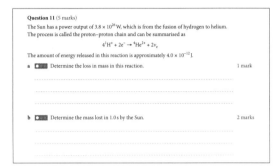

Practice exams

Both practice exams cover all content from Modules 5–8 of the HSC Physics syllabus. The practice exams have perforated pages so that you can remove them from the book and practise under exam-style conditions.

Solutions

Solutions to topic tests and practice exams are supplied at the back of the book. They have been written to reflect a high-scoring response and include marking guidelines and explanations of what makes an effective answer.

Explanations

The solutions section includes explanations of each multiple-choice option, both correct and incorrect. Explanations of the written response items explain what a high-scoring response looks like.

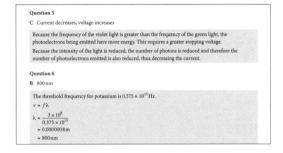
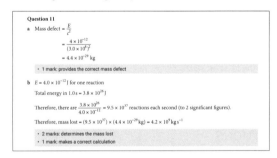

9780170465298

Icons

You will notice the following icons occurring in the summaries and exam practice sections of each chapter.

©NESA 2020 SI Q17

This icon appears with official past NESA questions.

These icons indicate whether the question is easy, medium or hard.

A+ HSC Physics Study Notes

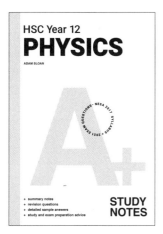

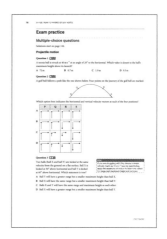

A+ HSC Physics Practice Exams can be used independently, or alongside the accompanying resource *A+ HSC Physics Study Notes*. *A+ HSC Physics Study Notes* includes topic summaries and exam practice for all key knowledge in the HSC Physics syllabus that you will be assessed on during the exam, as well as detailed revision and exam preparation advice to help you get ready for the exam.

A+ DIGITAL

Just scan the QR code or type the URL into your browser to access:

- A+ Flashcards: revise key terms and concepts online
- Revision summaries of all concepts from each inquiry question.

Note: You will need to create a free NelsonNet account.

https://get.ga/aplus-hsc-physics-u34

ABOUT THE AUTHORS

Paul Looyen

Paul Looyen is Head of Science and Agriculture at Macarthur Anglican School near Sydney. He has more than 28 years of experience teaching Physics. Paul has extensive marking experience of HSC external exams. He has participated on numerous CSSA Trial HSC examination committees for Physics. He has regularly presented at Meet the Markers run by the Science Teachers Association NSW. Paul has conducted HSC review student seminars via the University of Newcastle and the University of Wollongong. He has a popular YouTube channel, PhysicsHigh, to support students in understanding physics.

Sandra Woodward

Sandra Woodward has more than 20 years of experience teaching Science in several secondary schools. As a teacher of Science and Physics, Sandra is constantly searching for ways to inspire students and harness their curiosity, enthusiasm and energy to drive them to understand more about the sciences.

CHAPTER 1
MODULE 5: ADVANCED MECHANICS

Test 1: **Projectile motion**

Section I: 10 marks. Section II: 25 marks. Total marks: 35.
Suggested time: 60 minutes

Section I: Multiple-choice questions

Instructions to students
- For each question, circle the multiple-choice letter to indicate your answer.

Question 1

An aircraft is climbing at an angle and is accelerating both vertically and horizontally. A streamlined projectile is then dropped from the aircraft. Ignoring any air resistance, the acceleration of the projectile is

A less than g.

B equal to g.

C greater than g.

D zero.

Question 2 ©NESA 2021 SI Q1

A marble is rolled off a horizontal bench and falls to the floor.

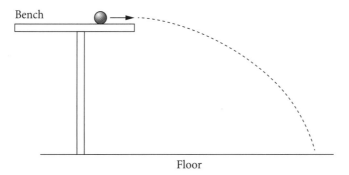

Rolling the marble at a slower speed would

A increase the range.

B decrease the range.

C increase the time of flight.

D decrease the time of flight.

Question 3 🔘🔘⬛

The diagram below shows the flight of a ball that was launched from position 1.

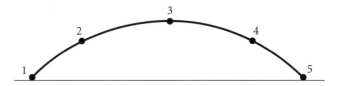

Which of the following statements is correct?

A The velocity at position 1 is the same as at position 5.

B The net force at position 3 is zero.

C The magnitude of the acceleration at position 2 is the same as at position 5.

D The horizontal velocity changes as the ball moves from left to right.

Question 4 🔘🔘⬛

Tom and Jeri each throw a small ball to try to hit a target that is 10 m away. They throw at the same time and the same speed, but Jeri throws the ball at a lower angle. They both hit the target.

Which of the following statements is correct?

A Jeri's ball will reach the target first because it has a greater acceleration.

B Jeri's ball will reach the target first because its vertical velocity will be less.

C Tom's ball will reach the target first because it achieves a greater height.

D Both balls will arrive at the same time because they have the same speed.

Question 5 🔘🔘⬛

A carnival bullseye is placed 10 m from a dart gun, as shown in the diagram below. The bullseye is set to fall the moment the dart gun is fired. The dart flies off at $40\,\mathrm{m\,s^{-1}}$.

In which direction should the dart gun be aimed to hit the bullseye?

A 31 cm above the bullseye

B At the bullseye

C 31 cm below the bullseye

D 1.2 m below the bullseye

Question 6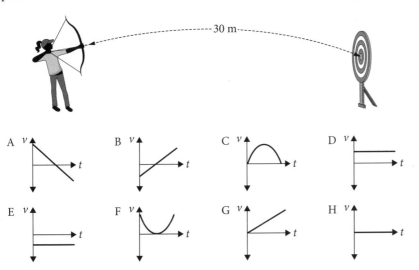

An arrow is shot upwards at an angle and lands at the same height 30 m away. Which pair of the graphs shown below represents the horizontal and vertical velocities of the arrow throughout the flight?

	Horizontal velocity	Vertical velocity
A	D	A
B	H	B
C	C	F
D	G	A

Use the information below to answer Questions 7 and 8.

During a spacewalk as part of the *Apollo 14* mission, Alan Shepherd hit a golf ball on the Moon. He then repeated the hit back on Earth, with the same angle and initial velocity. ($g_{\text{Moon}} = 1.62\,\text{m}\,\text{s}^{-2}$)

Question 7

Which of the following best describes how the maximum height and range of the golf ball would have differed when it was hit on Earth?

	Height	Range
A	Decreases	Decreases
B	Stays the same	Decreases
C	Decreases	Stays the same
D	Stays the same	Stays the same

Question 8

The times of flight were recorded for both hits. The ratio of the time of flight on the Moon to the time of flight on Earth is approximately

A 6:1

B 36:1

C 1:6

D 1:36

Question 9 ●●●

The paths of a projectile with and without air resistance are shown in the graph below.

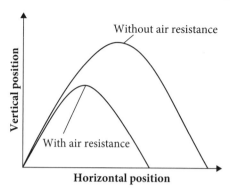

Air resistance causes the difference in the paths shown in the graph because it causes

A a diminishing force on the projectile.

B an increasing horizontal acceleration to the left.

C an increasing acceleration opposite to the direction of motion.

D a constant force to the left.

Question 10 ●●●

An object is launched at a velocity of $u \, \text{m s}^{-1}$ at an angle of $45°$. It lands at the same height as it was launched, R metres away. Which of the graphs below shows the correct relationship between the time of flight (t) and the range (R)?

A 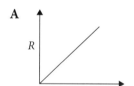   **B** 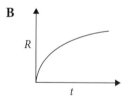   **C** 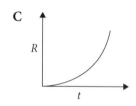   **D**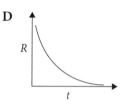

Section II: Short-answer questions

> **Instructions to students**
> •   Answer all questions in the spaces provided.

Question 11 (3 marks) ●○○

For a physics experiment, a student was (securely fastened) on the back of a truck that was moving at a constant $50 \, \text{km h}^{-1}$. The student threw a ball directly upwards into the air. Their lab partner videoed this from the side of the road so they could study the motion of the ball.

Explain the motion of the ball as seen by the lab partner taking the video.

Question 12 (4 marks)

A rower is moving across a 100 m river at a constant rate of 4.2 m s^{-1}. In the section of the river shown in the diagram below, the water is accelerating downstream at 2.0 m s^{-2}.

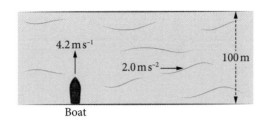

a ▢ Draw the path the boat will take.   1 mark

b ▢ Determine where the rower will land on the other bank.   3 marks

Question 13 (4 marks) ▢

A gazelle, trying to outrun a cheetah, reaches a river that is 4.0 m wide.

a What is the optimum angle at which the gazelle must jump to try to reach the other side of the river?   1 mark

b If the gazelle's maximum take-off speed is 6.0 m s^{-1}, determine whether it will make it to the other side of the river.   3 marks

Question 14 (5 marks)

Projectile motion can be modelled as an object following a parabolic path. However, this simplified model is only true if certain assumptions are made.

a Identify two assumptions that simplify projectile-motion analysis.   2 marks

b A projectile was launched on Earth at a velocity of $v \, \text{m s}^{-1}$ at an angle of θ, and its trajectory was tracked. Explain how its trajectory would differ if it were launched on the surface of the Moon with the same velocity and angle.   3 marks

Question 15 (5 marks)

A ball is launched horizontally at $5 \, \text{m s}^{-1}$ from a 50 m cliff. In the diagram below, the vertical gridline at $t = 0$ represents the cliff edge, and each horizontal gridline represents 5.0 m. Assume $g = 10 \, \text{m s}^{-2}$.

a On the grid mark the positions of the ball for the next 3 s.   3 marks

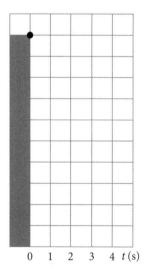

0   1   2   3   4 t (s)

b Determine the time when the ball will hit the ground.   2 marks

Question 16 (4 marks) ©NESA 2020 SII Q24 ●●◐

The graph shows the vertical displacement of a projectile throughout its trajectory. The range of the projectile is 130 m.

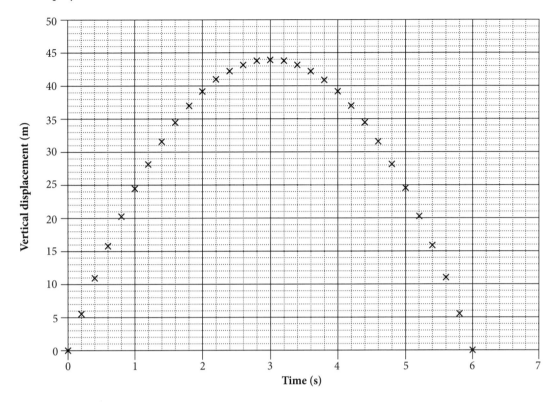

Calculate the initial velocity of the projectile.

Test 2: Circular motion

Section I: 10 marks. Section II: 25 marks. Total marks: 35.
Suggested time: 60 minutes

Section I: Multiple-choice questions

Instructions to students
- For each question, circle the multiple-choice letter to indicate your answer.

Question 1

Two ants are sitting on a spinning disc that has a 10 cm radius, as shown in the diagram.

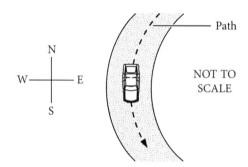

What is the ratio of their angular velocities, $\omega_A : \omega_B$?

A 1:1

B 2:1

C 1:2

D 5:10

Question 2 ©NESA 2000 SIA Q4

A car is moving at a *constant* speed along a circular path, as shown in the diagram.

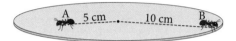

At the instant shown, the car is moving south. Which statement best describes the acceleration at this instant?

A The acceleration is directed to the south.

B The acceleration is directed to the west.

C The acceleration is directed to the east.

D The acceleration is zero.

Question 3

The diagram below shows a conical pendulum.

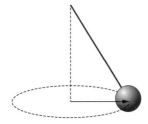

Which of the following options shows the net force(s) on the pendulum?

A   **B** ⟵   **C**   **D**

Question 4 ©NESA 2006 SI Q2

A mass attached to a length of string is moving in a circular path around a central point, O, on a flat, horizontal, frictionless table. This is depicted in the diagram below. The string breaks as the mass passes point X.

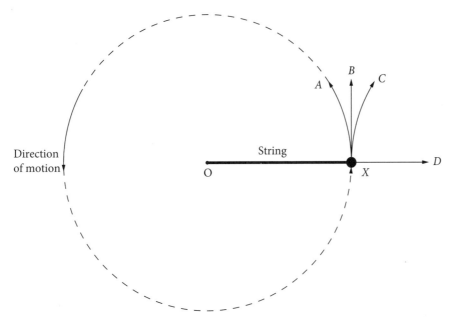

Which line best depicts the subsequent path of the mass?

A Line A

B Line B

C Line C

D Line D

Question 5 ●●●

The rotor at Luna Park 'sticks' people to the wall as it spins at a constant angular speed.

In which direction is friction acting?

A Up the wall

B Down the wall

C Into the circle

D Out of the circle

Use the information below to answer Questions 6 and 7.

A student carried out the following experiment in class. A string is passed through a tube. A mass is attached at each end. The student then swings the upper mass in a circular path.

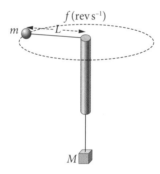

Question 6 ●●●

The student measured the radius, L, to be 1.5 m and the frequency to be 3.0 rev s^{-1}. Which of the following is the linear speed of mass m?

A 28 m s^{-1}

B 3.1 m s^{-1}

C 3.0 m s^{-1}

D 2.0 m s^{-1}

Question 7 ●●

Which of the following is an expression for the centripetal force?

A mf^2/L

B Mg

C $4\pi^2 rf$

D $4\pi^2 mL/M$

Question 8 ©NESA 2017 SI Q15 ●●●

A car travelling at a constant speed follows the path shown.

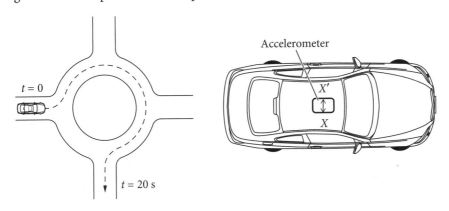

Accelerometer

$t = 0$

$t = 20$ s

X'

X

An accelerometer that measures acceleration along the X–X' direction is fixed in the car.

Which graph shows the measurements recorded by the accelerometer over the 20-second interval?

A

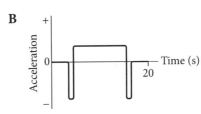

B

C

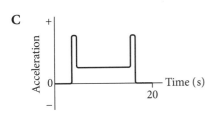

D

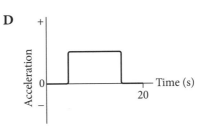

Question 9 ©NESA 2020 SI Q20 ●●●

The diagram shows a smooth, semi-circular, vertical wall with radius *r*.

A ball is launched from the position shown with a velocity *u* towards north at an angle to the horizontal.

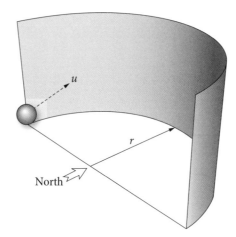

The ball follows a trajectory around the wall before landing on the ground, opposite its starting point. It does not reach the top of the wall.

Assume that there is no friction between the ball and the wall.

Which statement correctly describes the net force acting on the ball during its motion?

A The magnitude of the net force remains constant.

B The direction of the net force is vertically downwards.

C The direction of the net force is perpendicular to the wall.

D The magnitude of the net force reaches a minimum when the ball is at its highest point.

Question 10 ●●●

A person lifts a 100 kg mass at a constant rate using two pulleys, which are joined on their axle as shown. The mass is connected to the smaller pulley and the rope being pulled is connected to the larger pulley.

The closest value of the force the person needs to apply is

A 980 N

B 490 N

C 327 N

D 462 N

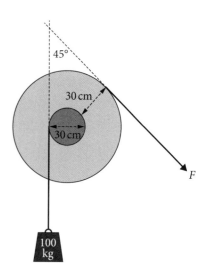

Section II: Short-answer questions

Instructions to students
Answer all questions in the spaces provided.

Question 11 (7 marks)

A record player turntable ($r = 15$ cm) turns at a rate of 45 rpm. A toy car is placed at the edge.

a What is the speed of the car?   2 marks

_____

_____

_____

_____

b If the car has a mass of 30 g, determine the frictional coefficient needed for the car to stay on the turntable.   3 marks

_____

_____

_____

_____

_____

_____

c The mass is now halved. How will this affect the car's ability to remain on the disc at the speed that was determined?   2 marks

_____

_____

_____

_____

Question 12 (5 marks)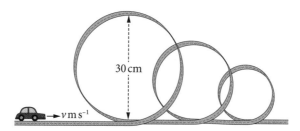

A toy car on a track travels around three vertical, consecutive loops, each smaller than the previous loop, as seen in the diagram below.

a Assuming negligible loss of energy from friction, what is the minimum speed the car must have so it can complete the first loop?

2 marks

b A student predicts that if the car completes the first loop, it will be successful in completing the remaining loops. Assess this prediction, with reference to physics principles and assumptions.

3 marks

Question 13 (4 marks)

One of the steepest banked curves in motor racing is at Winchester Speedway, in Indiana, USA, with a banking angle of 37° and a radius of turn of 250 m.

Assume the coefficient of friction between the tyres and the road is 0.70.

37°

a On the diagram above, draw and label the forces acting on the car when the car is travelling at its maximum possible speed.   2 marks

b The car (m = 1600 kg) approaches an ice patch that effectively reduces the friction to zero. What should its speed be to ensure the car stays on the track?   2 marks

---

Question 14 (9 marks)

A ball moves in a circular path at the end of a string, forming a conical pendulum, as shown in the diagram.

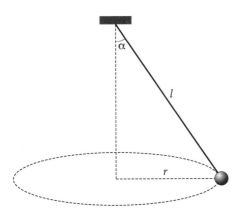

α

l

r

a  Draw and label the forces acting on the ball. Include the resultant force.   3 marks

b 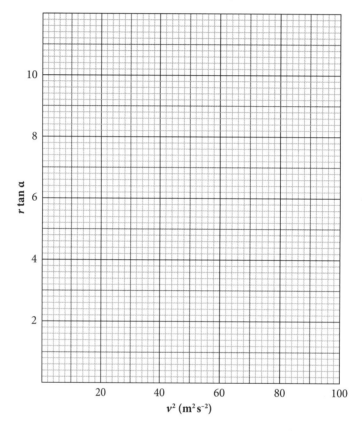 A student found that if the speed of the ball varies, the angle, α, changes. Some experimental values are recorded in the table below. Draw a v^2 versus $r \tan(\alpha)$ graph and draw a line of best fit.

3 marks

Radius, r (m)	Period (s)	Angle, α (°)	$\tan(\alpha)$	Velocity (m s^{-1})	Velocity^2 (m s^{-1})^2	$r \tan(\alpha)$
3.0	4.75	27.94	0.53	3.97	15.75	1.59
3.5	4.63	33.13	0.65	4.75	22.56	2.28
4.0	4.49	38.66	0.80	5.60	31.33	3.20
4.5	4.278	44.65	0.99	6.61	43.68	4.45
5.0	3.998	51.34	1.25	7.86	61.75	6.25
5.5	3.628	59.20	1.68	9.53	90.73	9.23

c Determine g using the gradient of the graph.

2 marks

d Assess the accuracy of the result.

1 mark

Test 3:   Motion in gravitational fields

Section I: 10 marks. Section II: 22 marks. Total marks: 32.
Suggested time: 60 minutes

Section I: Multiple-choice questions

Instructions to students
- For each question, circle the multiple-choice letter to indicate your answer.

Question 1 ©NESA 2003 SI Q2

A satellite moves in uniform circular motion around Earth. The following table shows the symbols used in the diagrams below. These diagrams are **not** drawn to scale.

Key

F	net force on satellite
v	velocity of satellite

Which diagram shows the direction of F and v at the position indicated?

A
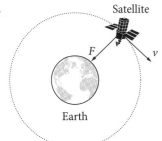

B
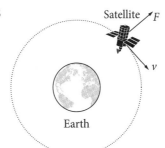

C
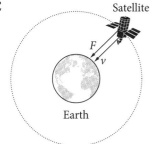

D
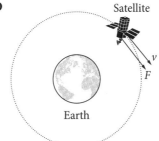

Question 2

The Hubble Space Telescope has an approximate mass of 11 000 kg, and its orbit is stable at an altitude of 547 km.

Its orbital velocity is approximately

A 3.4 km s^{-1}

B 6.3 km s^{-1}

C 7.6 km s^{-1}

D 9.6 km s^{-1}

Question 3 ⬤▢▢

An astronaut travels to a planet that has a gravitational field strength of $19.6 \, \text{N kg}^{-1}$ on its surface. If the planet has twice the mass of Earth, what is its radius?

A $6371 \, \text{km}$

B $12742 \, \text{km}$

C $3185 \, \text{km}$

D $1593 \, \text{km}$

Question 4 ⬤⬤▢

Two satellites, A and B, are in orbit around a planet that has a radius R. Their altitudes and masses are shown in the table below.

Satellite	Altitude	Mass
A	R	M
B	$2R$	$2M$

What is the ratio of the gravitational force each satellite experiences, $F_A : F_B$?

A $1:2$

B $2:3$

C $3:4$

D $9:8$

Question 5 ⬤⬤▢

A satellite orbits a planet in an elliptical orbit, as shown in the diagram below.

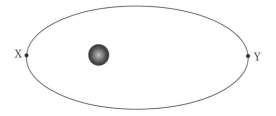

Which of the following statements is correct?

A The total energy at Y is greater than at X.

B The kinetic energy at Y is less than at X.

C The gravitational potential energy at Y is less than at X.

D The satellite will sweep out a greater area when it passes X than when it passes Y in the same time frame.

Question 6 ©NESA 2014 SIA Q6   ⬤⬤▢

A satellite is in a high orbit around the Earth. A particle of dust is in the same orbit.

Which row of the table correctly compares their potential energy and orbital speed?

	Potential energy	Orbital speed
A	Different	Same
B	Different	Different
C	Same	Same
D	Same	Different

Question 7

A satellite is moved from a low-Earth orbit to a geostationary orbit. Which of the graphs below best represents the rate of change in its gravitational potential energy?

A
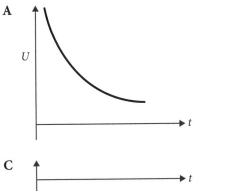

B
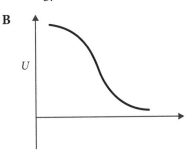

C
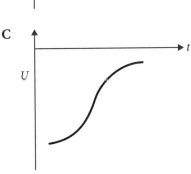

D
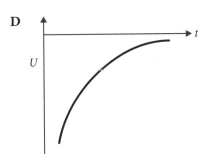

Question 8

A satellite moves between two planets, one larger than the other, as shown in the diagram below.

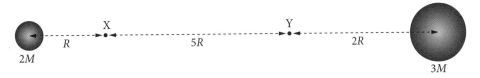

The distance between X and Y is $5R$ and the mass of the satellite is m.

As the satellite moves from X to Y, which of the following statements is true about the magnitude of the net force it experiences?

A The magnitude of the net force increases as the satellite moves from X to Y.

B The magnitude of the net force decreases as the satellite moves from X to Y.

C The magnitude of the net force increases and then decreases as the satellite moves from X to Y.

D The magnitude of the net force decreases and then increases as the satellite moves from X to Y.

Question 9

A space probe is launched to explore deep space. To do this, it must escape the gravitational pull of Earth. How could the escape velocity be reached for the minimum amount of fuel?

A Decrease the mass of the probe, thus increasing its acceleration.

B Increase the mass of the probe, thus increasing its inertia.

C Place it in orbit first before trying to attain the required velocity.

D Increase the force of the engines.

Question 10 ⬤⬤⬤

Two satellites are in unassisted orbit around two different planets. Satellite A orbits a planet of mass M at a distance of $2R$. Satellite B orbits a planet that is twice the first planet's mass and is in orbit at a distance of $3R$.

What is the ratio of their orbital velocities, $V_B : V_A$?

A 4 : 3

B 3 : 4

C 3 : 2

D $3 : 2\sqrt{2}$

Section II: Short-answer questions

> **Instructions to students**
> Answer all questions in the spaces provided.

Question 11 (3 marks) ⬤◯◯

Show that the mass of an orbiting satellite has no effect on the speed at which it orbits.

Question 12 (3 marks) ⬤⬤◯

The International Space Station circles Earth at an altitude of 408 km. Calculate the period at which it orbits to maintain this height.

Question 13 (4 marks) ●●▨

Astronauts in an orbiting spacecraft appear to be weightless. A student comments that this is because the force of gravity is non-existent in orbit about 300 km above Earth.

Assess this statement.

_____

_____

_____

_____

_____

_____

_____

_____

_____

Question 14 (3 marks) ●●▨

The James Webb Space Telescope (JWST) was placed in orbit at a point known as the L2 Lagrange point. A Lagrange point is a point at which the gravitational force from two large masses equals the centripetal force required for a small object to move with them.

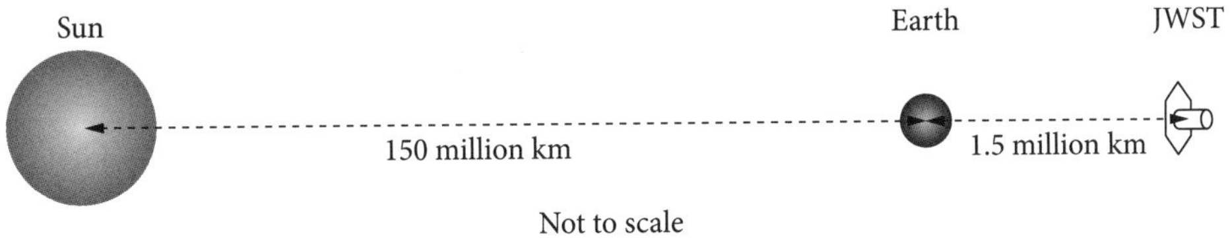

Sun   Earth   JWST

150 million km   1.5 million km

Not to scale

Determine the velocity of the JWST as it moves to this orbit. Assume the mass of the Sun is $M_S = 2.0 \times 10^{30}$ kg.

_____

_____

_____

_____

_____

_____

Question 15 (3 marks)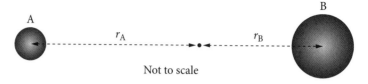

A 'binary' is a star system in which two stars rotate about a common point.

Not to scale

This common point is such that the ratio of the radii of the stars' orbits about this common point is equal to the ratio of their masses:

$$r_A : r_B = m_B : m_A$$

If star B is twice the mass of star A, and A has a period of 8 days, what is the period of B?

Question 16 (6 marks) ©NESA 2015 SII Q26

Consider the following two models used to calculate the work done when a 300 kg satellite is moved from Earth's surface to an altitude of 200 km.

You may assume that the calculations are correct.

Model X	Model Y
Data: g $= 9.8\,\mathrm{m\,s^{-2}}$ m $= 300\,\mathrm{kg}$ Δh $= 200\,\mathrm{km}$ $W = Fs$ $= mg\Delta h$ $= 3 \times 10^2 \times 9.8 \times 2.0 \times 10^5$ $= 5.9 \times 10^8\,\mathrm{J}$	Data: G $= 6.67 \times 10^{-11}\,\mathrm{N\,m^2\,kg^{-2}}$ r_{Earth} $= 6.38 \times 10^6\,\mathrm{m}$ r_{orbit} $= 6.58 \times 10^6\,\mathrm{m}$ M $= 6.0 \times 10^{24}\,\mathrm{kg}$ m $= 300\,\mathrm{kg}$ W $= \Delta E_p$ $\Delta E_p = E_{p\,\mathrm{final}} - E_{p\,\mathrm{initial}}$ $= -\dfrac{GMm}{r_{\mathrm{orbit}}} - \left(\dfrac{GMm}{r_{\mathrm{Earth}}}\right)$ $= -1.824 \times 10^{10} - (-1.881 \times 10^{10})$ $= 5.7 \times 10^8\,\mathrm{J}$

a ▉▉ What assumptions are made about Earth's gravitational field in models X and Y that lead to the different results shown?   2 marks

b ▉▉ Why do models X and Y produce results that, although different, are close in value?   1 mark

c ▉▉▉ Calculate the orbital velocity of the satellite in a circular orbit at the altitude of 200 km.   3 marks

CHAPTER 2
MODULE 6: ELECTROMAGNETISM

Test 4: Charged particles, conductors, and electric and magnetic fields

Section I: 10 marks. Section II: 26 marks. Total marks: 36.
Suggested time: 60 minutes

Section I: Multiple-choice questions

Instructions to students
- For each question, circle the multiple-choice letter to indicate your answer.

Question 1 ●○○

Calculate the electric field between two charged plates that are 30 mm apart, as shown in the diagram.

A $2.7\,\text{V m}^{-1}$

B $2700\,\text{V m}^{-1}$

C $2400\,\text{V m}^{-1}$

D $2.4\,\text{V m}^{-1}$

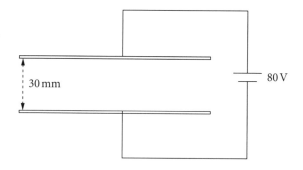

Use the information below to answer Questions 2 and 3.

Below are two plates, one positively charged and one negatively charged, with points identified between them.

Question 2 ○○○

The electric field at point *P* compared to the field at *O* is

A the same.

B greater.

C less.

D not possible to determine.

Question 3 ◯◯▮

The greatest amount of work would be done on a positive charge if it were moved from

A N to O.

B O to P.

C P to M.

D M to P.

Question 4 ◯◯▮

An electron is projected into a magnetic field with a velocity of $3.4 \times 10^2\,\mathrm{m\,s^{-1}}$.

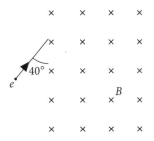

$B = 20\ \mu\mathrm{T}$

$v\ \mathrm{m\,s^{-1}}$

The magnitude and direction of the force on the charge when in the field is

A $1.09 \times 10^{-21}\,\mathrm{N}$ up the page.

B $1.09 \times 10^{-21}\,\mathrm{N}$ down the page.

C $1.09 \times 10^{-15}\,\mathrm{N}$ up the page.

D $1.09 \times 10^{-15}\,\mathrm{N}$ down the page.

Question 5 ©NESA 2008 SI Q11 ◯◯▮

An electron, e, moving with a velocity of $8.0 \times 10^6\,\mathrm{m\,s^{-1}}$ enters a uniform magnetic field, B, of strength $2.1 \times 10^{-2}\,\mathrm{T}$, as shown.

$40°$

e

B

The electron experiences a force which causes it to move along a circular path.

What is the radius of the path followed by the electron?

A $1.1 \times 10^{-3}\,\mathrm{m}$

B $1.4 \times 10^{-3}\,\mathrm{m}$

C $1.7 \times 10^{-3}\,\mathrm{m}$

D $2.2 \times 10^{-3}\,\mathrm{m}$

Question 6 🔘🔘⚫

An electron travels through a magnetic field, entering from the left as shown in the diagram below. The magnetic field strength is reduced as the electron moves from left to right.

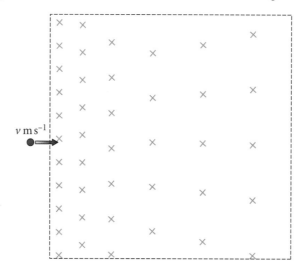

Which of the following best describes the trajectory of the electron?

A It will bend downwards with an increasing radius of curvature.

B It will bend downwards with a decreasing radius of curvature.

C It will bend upwards with an increasing radius of curvature.

D It will bend upwards with a decreasing radius of curvature.

Question 7 🔘🔘⚫

An alpha particle, made up of two protons and two neutrons, is fired through a region between electrically charged plates as shown.

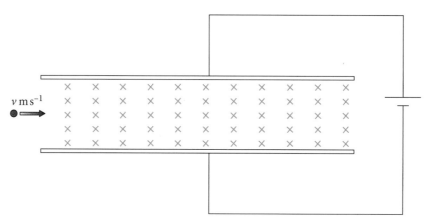

It passes through undeflected. If a proton were fired through at the same velocity, how would its motion change?

A The proton would pass through undeflected.

B The proton would follow a parabolic path downwards.

C The proton would follow a parabolic path upwards.

D The proton would speed up.

Question 8 ●●●

A positive charge is moving at a certain velocity, v. After some time, a uniform magnetic field is switched on. The magnetic field is large enough that the charge travels in a circular path and hence remains in the field, as shown in the diagram below.

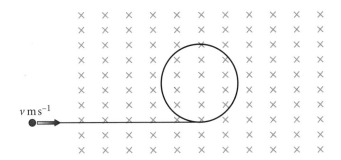

Which of the following variables would **not** affect its frequency of rotation?

A Charge, q

B Velocity, v

C Mass, m

D Magnetic field strength, B

Question 9 ●●●

An electron beam enters two electric fields that are perpendicular to each other, as shown in the diagram below.

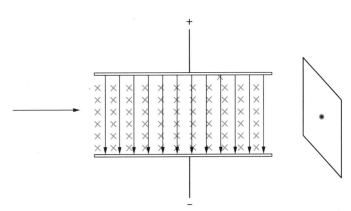

Initially, the fields are off and the beam is seen on a screen as a spot, as shown to the right of the diagram. The fields are turned on gradually. If the forces due to the fields have equal strengths, what path will the spot take, as seen from the left?

A 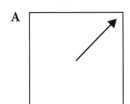   **B** 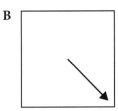   **C** 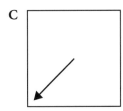   **D**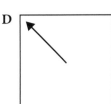

Question 10 ⬤⬤⬤

A charge enters a magnetic field at an angle, θ, as shown in the diagram below. Its motion is analysed for a set period of time.

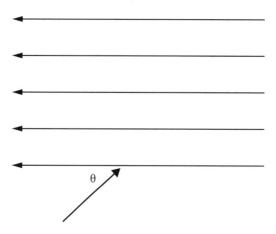

The angle is now reduced. What change occurs in its motion in the same period of time?

A It will move a lesser distance horizontally, with a larger radius of turn.

B It will move a greater distance horizontally, with a larger radius of turn.

C It will move a lesser distance horizontally, with a smaller radius of turn.

D It will move a greater distance horizontally, with a smaller radius of turn.

Section II: Short-answer questions

Instructions to students
- Answer all questions in the spaces provided.

Question 11 (4 marks) ⬤◐◐

For each of the scenarios in the table below, describe the behaviour of a positively charged object in the respective field.

Scenario	Electric field	Magnetic field
The charge is stationary.		
The charge enters the field at a constant velocity in the direction of the field lines.		
The charge enters the field at a constant velocity perpendicular to the field lines.		

Question 12 (6 marks)

An electron is accelerated between two charged plates, both of which have a hole to allow the electron to pass through, as shown in the diagram below.

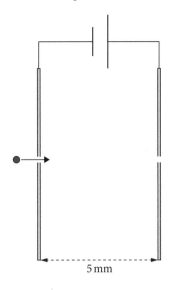

5 mm

a O Draw the electric field lines between the plates.   1 mark

b OO If the electric field strength is $3000\,\mathrm{N\,C^{-1}}$, determine the voltage between the plates.   2 marks

c OO What will be the velocity of the electron as it passes through the second hole, if it enters the first hole at $1.0 \times 10^7\,\mathrm{m\,s^{-1}}$?   3 marks

Question 13 (4 marks) ▮▮▯

Assess the energy changes of a child moving down a water slide as a model for a charge moving in an electric field.

_____

_____

_____

_____

_____

_____

_____

Question 14 (4 marks) ▮▮▯

A mass spectrometer can be used to compare the masses of different ions, as shown in the diagram below.

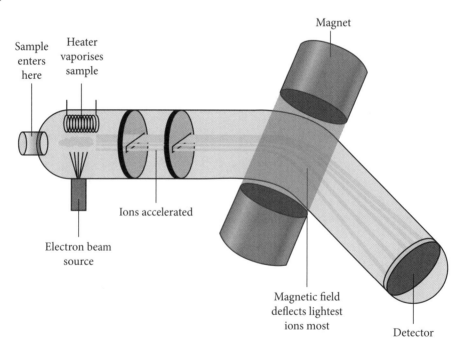

Explain this process, making reference to the diagram.

_____

_____

_____

_____

_____

_____

_____

_____

Question 15 (4 marks) ©NESA 2017 SII Q30 ●●●

In a thought experiment, a proton is travelling at a constant velocity in a vacuum with no field present. An electric field and a magnetic field are then turned on at the same time.

The fields are uniform in magnitude and direction and can be considered to extend infinitely. The velocity of the proton at the instant the fields were turned on is perpendicular to the fields.

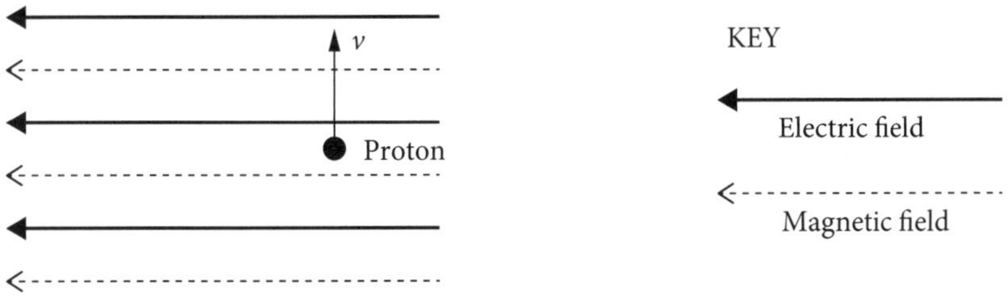

Analyse the motion of the proton after the fields have been turned on.

_____

_____

_____

_____

_____

_____

_____

_____

_____

Question 16 (4 marks) ⚫⚫⚫

An electron is fired at $1.0 \times 10^7 \, \text{m s}^{-1}$ between two 1.0 m plates, as shown in the diagram below.

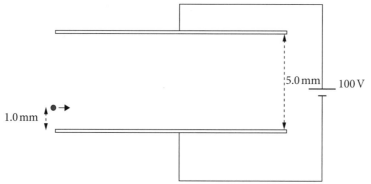

Neglecting gravitational forces, determine how far the electron will travel into the field.

_____

_____

_____

_____

_____

_____

_____

_____

Test 5: The motor effect

Section I: 10 marks. Section II: 25 marks. Total marks: 35.
Suggested time: 60 minutes

Section I: Multiple-choice questions

Instructions to students
- For each question, circle the multiple-choice letter to indicate your answer.

Question 1

Determine the force on the current-carrying wire placed in the magnetic field shown in the diagram below.

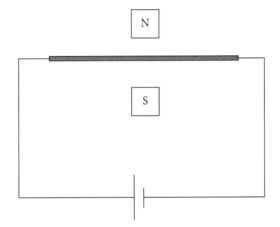

A Up

B Down

C Into the page

D Out of the page

Question 2

A current-carrying wire is placed in a magnetic field, as shown in the diagram below.

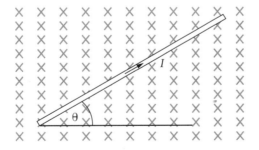

What happens when the angle θ is increased?

A The force on the wire increases.

B The force on the wire decreases.

C The force on the wire remains the same.

D The current decreases.

Use the information below to answer Questions 3 and 4.

A 50 g rod is held horizontally and allowed to fall vertically under the effect of gravity, while it is still connected to a power supply, through a magnetic field of 0.2 T. When the current is turned on, the rod falls at a constant rate.

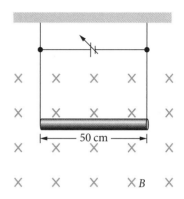

Question 3

Determine the current that is needed for the rod to fall at a constant rate.

A 4.9 A left to right

B 4.9 A right to left

C 49 A left to right

D 49 A right to left

Question 4

A student discovers that for a certain magnetic field and current the rod accelerates upwards. What changes to the experiment can the student make so that the rod has no acceleration?

A Increase the length of the rod.

B Increase the magnetic field.

C Decrease the current.

D Decrease the mass.

Question 5

Determine the magnitude and direction of the force on the wire in a magnetic field shown in the diagram.

A 5.0 N left

B 1.7 N left

C 5.0 N right

D 1.7 N right

Question 6 ©NESA 2018 SIA Q5

The diagram shows a current-carrying conductor in the magnetic field.

What is the magnitude of the force on the conductor?

A 0 N

B 0.05 N

C 0.09 N

D 0.10 N

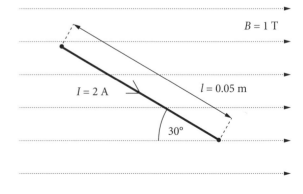

Question 7 ◐▨▨

Two current-carrying wires are placed next to each other and the force of repulsion between them is measured to be $1.3 \times 10^{-4}\,\mathrm{N\,m^{-1}}$. If the current in the lower wire is 2.0 A to the right, what is the magnitude and direction of the current in the upper wire?

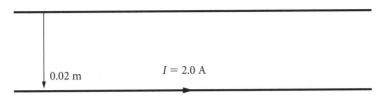

A 6.5 A left to right

B 6.5 A right to left

C 2 A right to left

D 2 A left to right

Question 8 ▨▨▨

A teacher wants to determine the strength of a big magnet that was found in the storeroom of the science department. She sets up the equipment as shown in the diagram below.

The power supply is set to 10 V and a rheostat is used to control the current in the circuit.

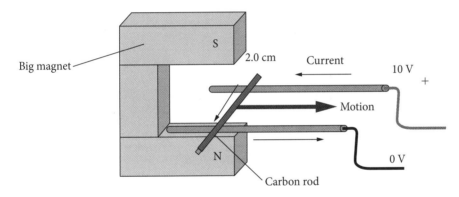

The carbon rod has a mass of 10 g and sits on two carbon rods that are connected to the circuit, raised at 20°. 2.0 cm of the rod sits within the magnetic field.

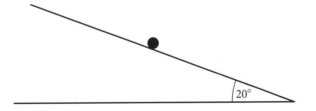

The rheostat is adjusted so that the rod moves at a constant speed. At this point the resistance is 6.25 Ω.

Calculate the magnetic field strength of the big magnet.

A 1.1 T

B 3.1 T

C 0.18 T

D 1.0 T

Question 9 ©NESA 2017 SIA Q13 ○○○

A triangular piece of wire is placed in a magnetic field as shown.

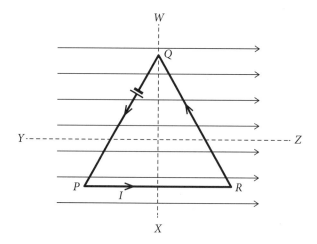

When current I is supplied as shown, how does the wire move?

	Axis of rotation	Direction of movement
A	YZ	Q into page
B	YZ	Q out of page
C	WX	R into page
D	WX	R out of page

Question 10 ○○○

The diagram below shows a cross-section of a loudspeaker, in which a coil is placed over a cone. The cone is then free to move to generate sound.

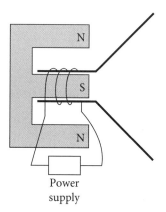

Which of the following would produce the loudest sound?

A DC supply with increased coil resistance

B DC supply with decreased coil resistance

C AC supply with increased coil resistance

D AC supply with decreased coil resistance

Section II: Short-answer questions

Instructions to students
• Answer all questions in the spaces provided.

Question 11 (4 marks) ©NESA 2013 SIB Q25

P, Q and R are straight current-carrying conductors.

Conductors P and R are fixed and unable to move. Conductor Q is free to move.

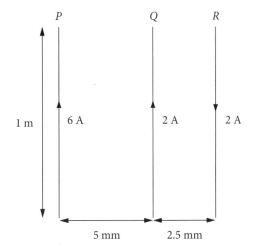

a ●●▨ In which direction will the conductor Q move as a result of the current flow in P and R?

1 mark

b ●●▨ Calculate the magnitude of the force experienced by Q as a result of the currents through P and R.

3 marks

Question 12 (3 marks)

Two current-carrying rods are placed close together, as shown in the diagram. They both experience a force.

Explain what happens. In your answer, make reference to the motor effect and the relevance to Newton's laws.

_____

_____

_____

_____

_____

_____

Question 13 (9 marks)

An experiment is carried out to determine the value of the magnetic permeability constant, μ_0. This is done by measuring the force between 2.0 m-long current-carrying wires, both with a current of 2.0 A, as they are moved different distances apart.

After many trials, the following average results were obtained.

Force, F ($\times 10^{-5}$ N)	Distance, d (m)
8.0	0.01
4.0	0.02
2.7	0.03
2.0	0.04
1.6	0.05

a   Plot a graph to show a linear relationship that would allow the calculation of μ_0.   3 marks

b ⚫⚫⬜ Use the data from your graph to determine μ_0.   3 marks

_____

_____

_____

_____

_____

_____

c ⚫⚫⚫ Compare your value from part **b** with the value on the data sheet. Propose a reason
(or reasons) for any discrepancy.   3 marks

_____

_____

_____

_____

_____

_____

Question 14 (4 marks)

Two conductors are connected to circuits as shown in the diagram below.

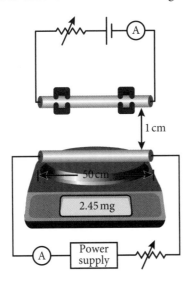

The top rod cannot move and carries a constant current. The lower wire lies on the electronic
balance with the current switched off. The balance is 'zeroed' so that the display shows a
reading of zero mass. When the power supply is switched on and a current of 1.2 A flows
through the wire, the balance reads 2.45 mg.

a ⚪⬜⬜ In which direction does the current flow in the rod for the reading on the balance
to be as stated above?   1 mark

_____

_____

b ⬤◯◯ Determine the value of the force that must be generated to obtain the reading on the scale.   1 mark

_____

_____

c ⬤⬤◯ Calculate the current needed in the rod to generate this force.   2 marks

_____

_____

_____

_____

Question 15 (5 marks) ⬤⬤◯

A 5.0 cm rod of mass 10 g is suspended between two magnets by two light wires of negligible mass, as shown in the diagram below.

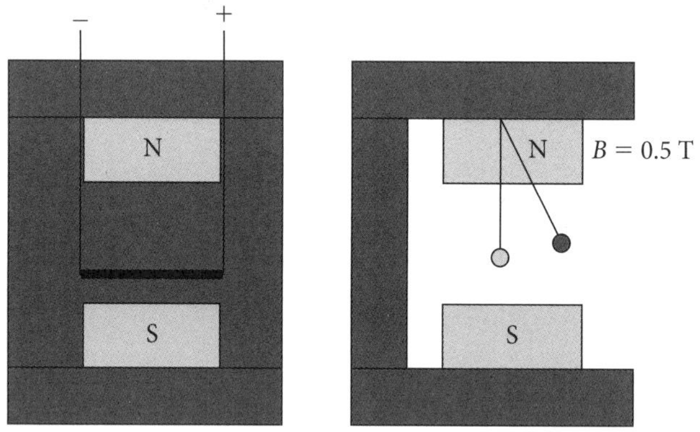

When a current of 1.5 A is applied, the rod swings outwards at an angle.

a What is the magnitude of the electromagnetic force experienced by the rod?   2 marks

_____

_____

_____

_____

b Determine the angle at which it stops.   3 marks

_____

_____

_____

_____

_____

_____

Test 6: Electromagnetic induction

Section I: 10 marks. Section II: 25 marks. Total marks: 35.
Suggested time: 60 minutes

Section I: Multiple-choice questions

Instructions to students
- For each question, circle the multiple-choice letter to indicate your answer.

Question 1

The diagram below shows a magnetic field B passing through a loop of area A at an angle θ to the normal.

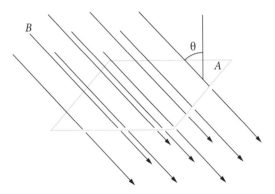

Which of the following changes will increase the flux through the loop?

A Decrease A

B Increase θ to $90°$

C Increase B

D Decrease B

Question 2

Calculate the flux passing through the circular loop shown in the diagram below.

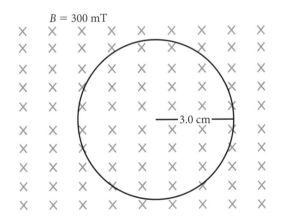

A $8.5\,\text{Wb}$

B $2.7 \times 10^{-4}\,\text{Wb}$

C $8.5 \times 10^{-4}\,\text{Wb}$

D $2.8 \times 10^{-2}\,\text{Wb}$

Question 3 ⬤◼◼

Calculate the voltage in the secondary coil of the transformer shown below.

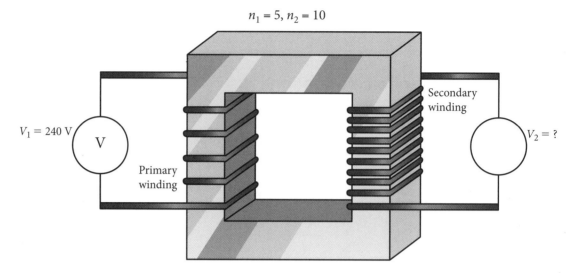

$n_1 = 5, n_2 = 10$

$V_1 = 240$ V

Primary winding

Secondary winding

$V_2 = ?$

A 60 V

B 120 V

C 240 V

D 480 V

Question 4 ⬤◼◼

A student designs a simple step-down transformer with 100 windings around an iron nail and 50 windings around a secondary nail, as shown in the diagram below.

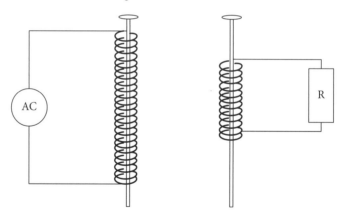

How can the efficiency of this transformer be improved?

A Increase the length of the windings.

B Wind both coils onto the same nail.

C Increase the number of turns in both coils, keeping the ratio the same.

D Replace the iron nails with more conductive copper pegs.

Question 5 ©NESA 2005 SIA Q10 ⬤⬤

A transformer is to be designed so that it is efficient, with heating by eddy currents minimised. The designer has some iron and insulating material available to build the transformer core. The windings are to be made with insulated copper wire.

Which of the following designs minimise the energy losses in the core?

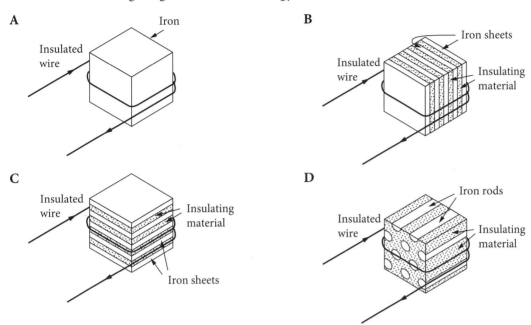

Question 6 ⬤⬤

A magnet is dropped down a copper pipe, as shown in the diagram.

As the magnet falls, eddy currents are produced. What is the direction of the eddy currents produced at positions A and B, as seen from above?

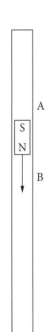

	Position A	Position B
A	Clockwise	Clockwise
B	Clockwise	Anticlockwise
C	Anticlockwise	Clockwise
D	Anticlockwise	Anticlockwise

Question 7 ©NESA 2013 SIA Q13 ●●

Different magnetic fields are passing through two copper rings, *P* and *Q*, as shown.

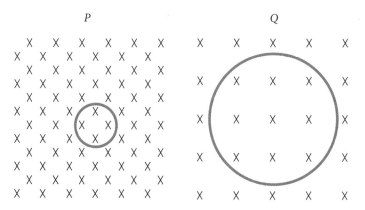

Which row of the table correctly identifies the ring with the greater magnetic flux and the ring with the greater magnetic flux density?

	Greater magnetic flux	**Greater magnetic flux density**
A	*P*	*P*
B	*Q*	*Q*
C	*P*	*Q*
D	*Q*	*P*

Question 8 ●●●

A circular loop of wire is placed in a magnetic field and rotated in the direction shown in the diagram.

Which statement is true for the current position of the wire?

A The current is zero.

B The current is at a maximum and is in a clockwise direction.

C The current is at a maximum and is in an anticlockwise direction.

D The rate of change of flux produces an alternating current.

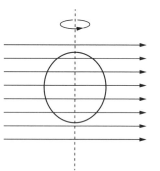

Question 9 ●●

A rare-earth magnet and a copper block are suspended by strings from a fixed rod, as shown in the diagram. The copper is pulled back and allowed to swing towards the magnet.

Which of the following statements best describes the resulting motion of the magnet?

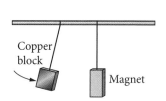

A The copper sticks to the magnet and they move off together.

B The copper slows suddenly and stops before reaching the stationary magnet.

C The copper repels the magnet and then attracts it in a continued oscillating motion.

D Because copper isn't attracted by a magnet, they continue to hit each other and move away, as in a Newton's cradle.

Question 10 ⬤⬤⬤

A coil of wire was moved through a magnetic field, from a to e, at a constant rate, as shown in the diagram below.

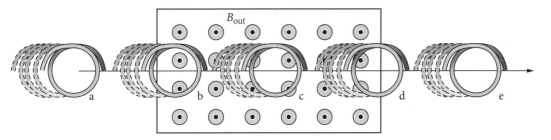

A graph was drawn of the EMF generated over time. Positive is defined as clockwise. Which of the following graphs best represents what is happening?

A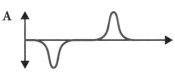

B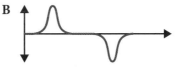

C

D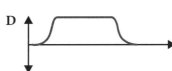

Section II: Short-answer questions

Instructions to students
• Answer all questions in the spaces provided.

Question 11 (5 marks)

A square loop of wire of sides 10 cm is placed in a magnetic field as shown in the diagram below.

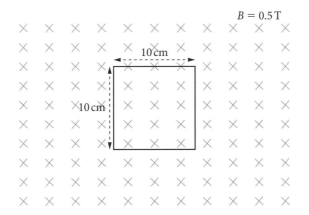

a ⬤⬜⬜ Determine the flux passing through the loop. 2 marks

b 🔵🔵⬜ The wire is now rotated 45° in the direction shown in the diagram below, in 0.5 s.   3 marks

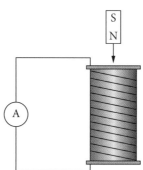

What is the EMF generated?

_____

_____

_____

_____

_____

_____

Question 12 (6 marks) 🔵🔵⬜

A magnet is dropped into a coil that is connected to an ammeter, as shown in the diagram below.

The current generated is recorded. A graph of current versus time is shown below.

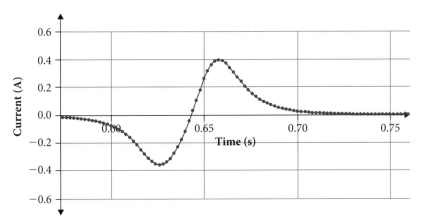

a Explain why the second peak of the graph has a larger amplitude than the first peak.   2 marks

_____

_____

_____

_____

b The experiment is repeated, but this time the magnet enters the coil at a faster speed.
Sketch the curve of current versus time for this new speed.   2 marks

c If the magnet polarity was reversed, would anything change? Explain your answer.   2 marks

_____

_____

_____

_____

9780170465298

Question 13 (5 marks)

Students wanted to see how to make a transformer with the highest possible efficiency. They measured the voltage and current in both the primary and secondary coils in the four set-ups shown below.

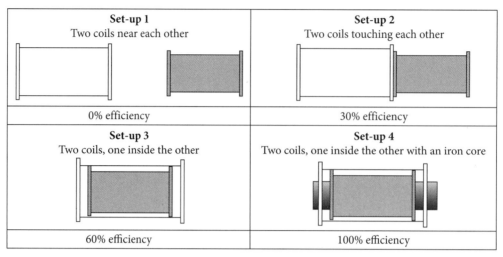

Set-up 1 Two coils near each other	Set-up 2 Two coils touching each other
0% efficiency	30% efficiency
Set-up 3 Two coils, one inside the other	Set-up 4 Two coils, one inside the other with an iron core
60% efficiency	100% efficiency

Referring to the results, explain the best arrangement for a highly efficient transformer.

Question 14 (4 marks)

Electricity is generated at power stations and transmitted at voltages much higher than that needed in the home, as shown in the diagram below.

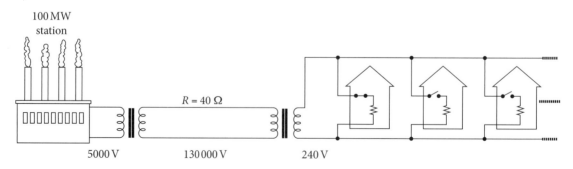

100 MW station

$R = 40\ \Omega$

5000 V   130 000 V   240 V

NESA 2018 SII Q30 (ADAPTED)

Explain the role of transformers in getting electricity to homes safely.

_____

_____

_____

_____

_____

_____

_____

_____

_____

Question 15 (5 marks) ©NESA 2017 SIB Q27

The diagram shows an electric circuit in a magnetic field directed into the page. The graph shows how the flux through the conductive loop changes over a period of 12 seconds.

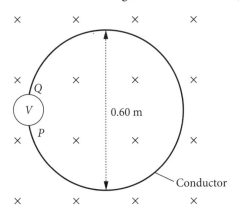
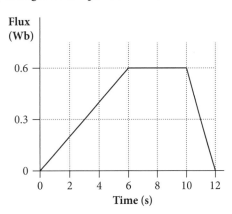

a ●● Calculate the maximum magnetic field strength within the stationary loop during the 12-second interval.   2 marks

_____

_____

_____

_____

b ●●● Calculate the maximum voltage generated in the circuit by the changing flux. In your answer, indicate the polarity of the terminals P and Q when this occurs.   3 marks

_____

_____

_____

_____

_____

_____

Test 7: Applications of the motor effect

Section I: 10 marks. Section II: 16 marks. Total marks: 26.
Suggested time: 50 minutes

Section I: Multiple-choice questions

Instructions to students
- For each question, circle the multiple-choice letter to indicate your answer.

Question 1

Below are diagrams of AC and DC generators.

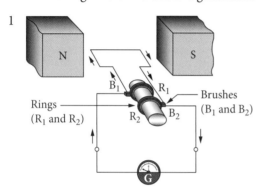
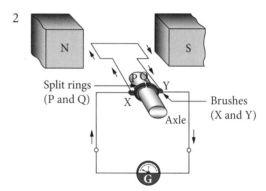

Graphs of output voltage versus time are shown below.

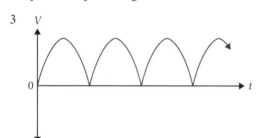
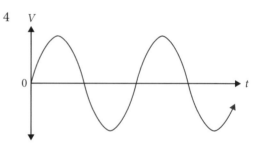

Which of the following options correctly matches the diagrams to the graphs of output voltage?

A 1 matches to 4, 2 matches to 3.

B 1 matches to 3, 2 matches to 4.

C 1 matches to 5, 2 matches to 3.

D 1 matches to 5, 2 matches to 4.

Question 2 ⬤⬤⬤

Which statement is true about an induction motor?

A The stator produces a rotating magnetic field.

B It has brushes to maintain contact with the commutator.

C The rotor receives its current from an external supply.

D It has a fixed rate of rotation dependent on its AC supply.

Question 3 ⬤⬤⬤

The following graph shows the current in a drill as it is turned on.

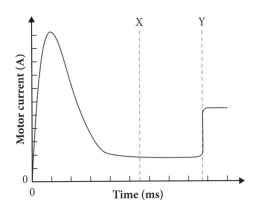

Which of the following best describes what is happening at X and Y?

	X	Y
A	Back EMF at maximum	Load applied
B	Back EMF at minimum	Load applied
C	Back EMF at maximum	Load removed
D	Back EMF at minimum	Load removed

Question 4 ⬤⬤⬤

A simple electric motor is shown in the diagram below.

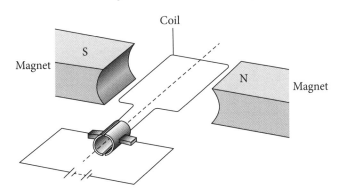

Determine the direction the coil will turn (looking along the axis of rotation into the page).

A Clockwise

B Anticlockwise

C The coil will not turn, but will move backwards and forwards.

D The coil will not turn.

Use the information below to answer Questions 5 and 6.

A student sets up a current balance as shown in the diagram below. A current-carrying loop of wire experiences two forces. One is the force due to gravity on the hanging mass; the other is from a magnetic field from the magnet to the left.

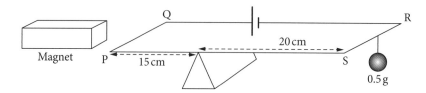

Question 5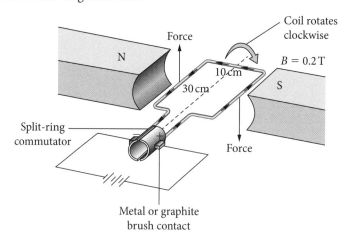

What is the direction of the current and the magnetic field at PQ to balance the mass at RS?

	Current	Magnetic field
A	Into the page	Right to left
B	Into the page	Left to right
C	Out of the page	Right to left
D	Out of the page	Left to right

Question 6

If the magnetic field at PQ is 5 mT and the lengths PQ and RS are 10 cm, what current must be applied to balance the loop?

A 0.13 A

B 13.1 A

C 9.8 A

D 0.098 A

Question 7

A simple motor is shown in the diagram below.

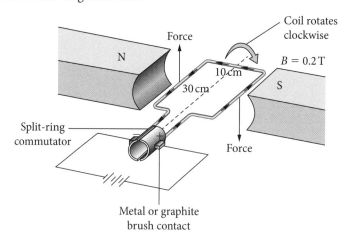

For the following data, determine the torque generated in the motor.

$$B = 0.2 \text{ T}, V = 10 \text{ V}, R = 3\,\Omega, n = 10 \text{ turns}$$

A 0.2 N m

B 0.03 N m

C 0.6 N m

D 2000 N m

Question 8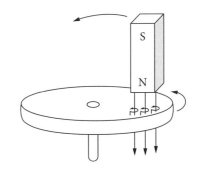

The diagram shows an experiment students carried out in class to simulate the workings of an AC induction motor.

Which of the following correctly matches the experiment part to its corresponding AC motor part?

	Experiment part	AC induction motor part
A	Disc	Squirrel cage
B	Magnet	Rotor
C	Disc	Stator
D	Magnet	Squirrel cage

Question 9

Magnets can be used for braking. In the process, heat is generated, as seen in the heat image.

This can best be explained by which of the following statements?

A Heat always results from eddy currents.

B Conservation of energy sees the loss in kinetic energy result in the production of heat energy.

C Heat is generated by friction from the disc and magnets rubbing together.

D The magnetic field generates an opposing force to slow the disc.

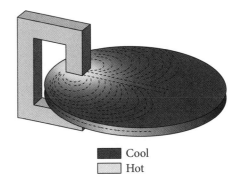

■ Cool
□ Hot

Question 10 ©NESA 2013 SIA Q15 (ADAPTED)

The diagram shows a single-loop motor.

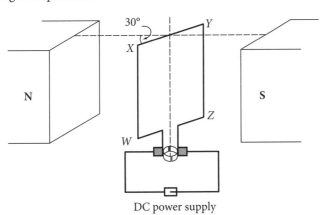

DC power supply

The equations $\tau = nBIA \sin\theta$ and $F = BIl \sin\omega$ can be used to calculate the torque in the motor and the force on the length of wire WX, respectively.

What angles are represented by θ and ω in the above equations?

	θ	ω
A	30°	90°
B	30°	30°
C	60°	90°
D	60°	30°

Section II: Short-answer questions

> **Instructions to students**
> - Answer all questions in the spaces provided.

Question 11 (3 marks)   ©NESA   2004 SIB Q21a

The diagram shows a two-pole DC motor as constructed by a student.

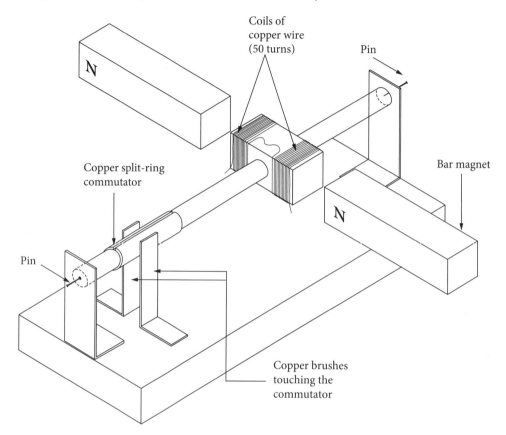

Identify **three** mistakes in the construction of this DC motor as shown in the diagram.

Question 12 (3 marks)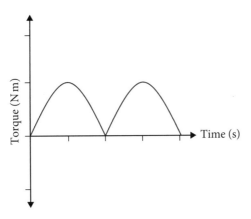

Back EMF decreases the current flowing in a circuit. Describe how it is still possible for
a motor to spin.

_____

_____

_____

_____

_____

_____

Question 13 (3 marks)

A student designs a simple DC motor and measures the torque generated for 1 s. The results
are shown in the graph below.

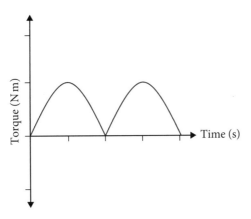

Discuss how the position of the coil relates to the results shown in the graph.

_____

_____

_____

_____

_____

_____

Question 14 (4 marks) ⬤⬤⬤

The Hair Raiser at Luna Park, shown here, drops a ring of seated people from a height of 50 m. The ring slows suddenly, almost stopping, through magnetic braking, and it then comes to a stop as it reaches the ground.

A graph of the velocity squared versus distance from the top of the ride is shown below.

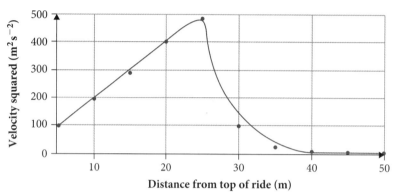

Explain how this motion is achieved, with reference to the graph.

_____

_____

_____

_____

_____

_____

_____

Question 15 (3 marks) ⬤⬤⬤

A student designs a simple DC motor with a set length of wire, and constructs a circular loop. After measuring the maximum torque, they repeat the experiment using the same wire but doubling the radius. Using quantitative analysis, show that the maximum torque is now double the initial value.

_____

_____

_____

_____

_____

CHAPTER 3
MODULE 7: THE NATURE OF LIGHT

Test 8: Electromagnetic spectrum

Section I: 5 marks. Section II: 15 marks. Total marks: 20.
Suggested time: 30 minutes

Section I: Multiple-choice questions

Instructions to students
- For each question, circle the multiple-choice letter to indicate your answer.

Question 1

James Clerk Maxwell is best known for

A measuring the speed of electromagnetic waves.

B unifying the laws of electricity and magnetism.

C determining the speed of light.

D proving the existence of electromagnetic waves.

Question 2

A student observed spectra that were obtained from sources of light, but forgot to label them. The images they recorded are shown below.

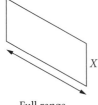

Full range
of colours

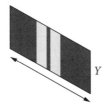

Two yellow lines on
a dark background

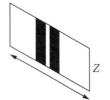

Range of colours
with two black lines

Which of the following labels correctly match the spectra?

	X	**Y**	**Z**
A	Incandescent bulb	Star	Discharge tube
B	Discharge tube	Incandescent bulb	Star
C	Star	Discharge tube	Incandescent bulb
D	Incandescent bulb	Discharge tube	Star

Question 3 ●●

An image of the spectrum of a star along with the spectra of elements as measured in a laboratory are shown below.

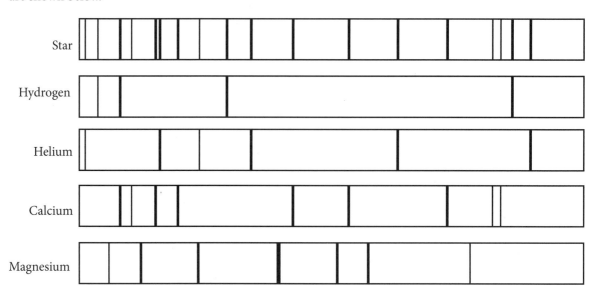

Determine which element is **not** present in the atmosphere of this star.

A H

B He

C Ca

D Mg

Question 4 ●●●

In 1862, Jean Bernard Léon Foucault determined the speed of light by shining a light on a rotating mirror, which reflected light onto a fixed mirror, as shown in the diagram below.

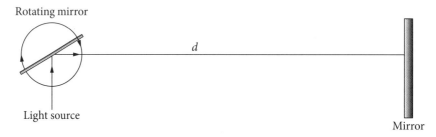

By the time the light had returned to the rotating mirror, the mirror had rotated slightly from position 1 to position 2, as shown in the diagram below. Foucault was able to measure the speed of light by measuring the angle between the light source and the angle of the reflected light to determine the angle between position 1 and position 2.

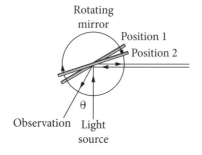

Assuming the mirror had an angular velocity of ω rad s^{-1}, which of the following expressions would allow Foucault to determine the speed of light, c?

A $\dfrac{2d\omega}{\theta}$

B $\dfrac{d\omega}{\theta}$

C $\dfrac{4d\omega}{\theta}$

D $\dfrac{d\omega}{2\theta}$

Question 5

The diagram shows two absorption spectra: one of hydrogen, measured in a laboratory, and the other from a star.

Which of the following **cannot** be determined about the star from this information?

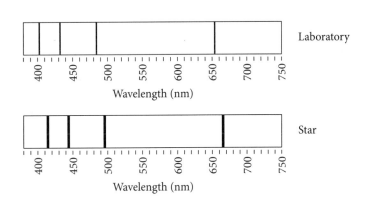

A Chemical composition

B Rotational velocity

C Translational velocity

D Surface temperature

Section II: Short-answer questions

Instructions to students
- Answer all questions in the spaces provided.

Question 6 (3 marks)

Two sources of light, X and Y, produce different types of spectra. Explain the difference between the spectra produced by X and Y.

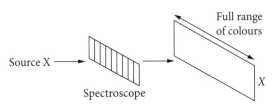

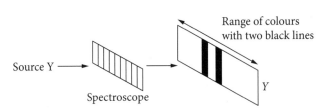

Question 7 (3 marks)

Compare the current method used to determine the speed of light with a historical method.

_____

_____

_____

_____

_____

_____

Question 8 (5 marks)

Maxwell's theories developed a mathematical relationship between electric and magnetic fields. This unified previous theories in physics. Briefly discuss the implications and importance of this and recount evidence supporting his theory.

_____

_____

_____

_____

_____

_____

_____

_____

_____

Question 9 (4 marks)

The image below shows a spectrum of a distant star.

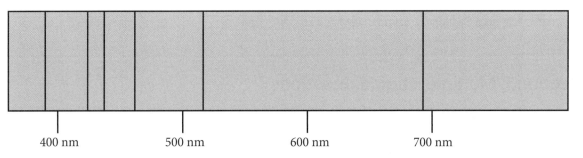

400 nm   500 nm   600 nm   700 nm

The absorption lines were matched to the emission spectrum of hydrogen, which is shown below.

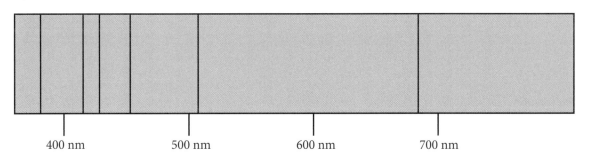

400 nm   500 nm   600 nm   700 nm

Explain what can be determined about the star from these spectra.

_____

_____

_____

_____

_____

_____

_____

_____

Test 9: Light: wave model

Section I: 10 marks. Section II: 25 marks. Total marks: 35.
Suggested time: 60 minutes

Section I: Multiple-choice questions

Instructions to students
- For each question, circle the multiple-choice letter to indicate your answer.

Question 1

Which characteristic of waves is represented in the diagram?

A Interference

B Diffraction

C Reflection

D Refraction

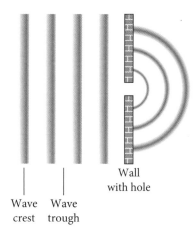

Question 2

Determine the wavelength of light that produced the pattern shown in the diagram below, where $w = 2\,mm$.

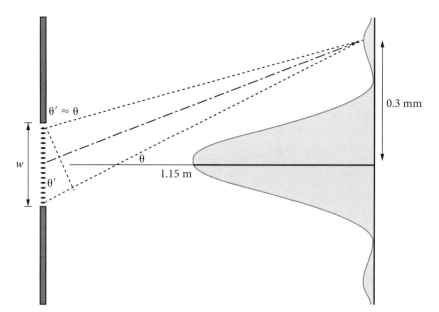

A 0.52 mm

B 5.22 nm

C 522 nm

D 0.17 mm

Use the diagram below to answer Questions 3 and 4.

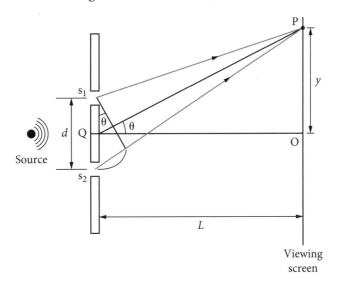

Question 3

Laser light of 640 nm is incident on a slit of 0.30 mm and produces an interference pattern with the distance to the first maxima $y = 5.0$ cm.

The value of L, the distance to the screen, is closest to

A 2.3×10^{-3} m

B 430 m

C 4.2×10^{-3} m

D 23 m

Question 4

The distance to the screen, L, is now halved. What would be the new distance to the first maxima?

A 0.1 m

B 0.4 m

C 0.025 m

D 0.15 m

Question 5

Which of the following statements is true about Newton's and Huygens' models of light?

	Newton	Huygens
A	Light travels as particles.	Light travels as transverse waves.
B	Light travels as transverse waves.	Light travels as longitudinal waves.
C	Light travels as particles.	Light travels as longitudinal waves.
D	Light travels as transverse waves.	Light travels as particles.

Question 6 〇〇▪

Light passes through two polarising filters. The second filter's axis is at an angle of 30° to the first. What percentage of the light that entered the filters will exit them?

A 50%

B 86%

C 75%

D 38%

Question 7 〇〇▪

Light with a wavelength of 680 nm is shone through a 10 μm double slit onto a screen 1 m away. What is the maximum possible number of maxima?

A 14

B 15

C 14.7

D 1.4

Question 8 〇〇〇

Two polarisers are set up as shown in the diagram.

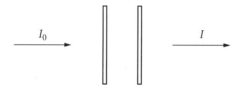

If the angle between the two filters is θ and I is one-third of I_0, then θ is closest to

A 71°

B 55°

C 35°

D 66°

Question 9 〇〇〇

The following statements refer to the strengths and weaknesses of Newton's and Huygens' models of light.

　I Newton's model explained refraction by an increase in the speed of light, whereas Huygens' model explained refraction by a decrease in the speed of light.

　II Newton's model explained refraction by a decrease in the speed of light, whereas Huygens' model explained refraction by an increase in the speed of light.

　III Neither model could explain polarisation.

　IV Newton's model could not explain polarisation, whereas Huygens' model could.

Which statements are correct?

A I, III

B I, IV

C II, III

D II, IV

Question 10 ⬤⬤⬤

A student sets up a double-slit experiment with a known wavelength of light, λ. She records the distance to the first maxima, y, on a screen that is L metres away. She then changes the wavelength of the light and moves the screen so that y remains constant.

Which graph represents the relationship between λ and L?

A   **B** 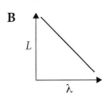   **C** 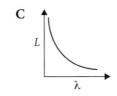   **D**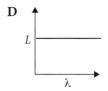

Section II: Short-answer questions

> **Instructions to students**
> • Answer all questions in the spaces provided.

Question 11 (6 marks) ⬤◐◐

Newton and Huygens suggested models of light as they attempted to understand its nature.

a Discuss an observation that allowed them to come up with their respective models.   4 marks

_____

_____

_____

_____

_____

_____

_____

_____

b Models are used to help understand phenomena but have limitations. Identify a limitation of each of these models.   2 marks

_____

_____

_____

_____

Question 12 (4 marks)

A student studying Malus' law recorded how the intensity of light changes with the angle between the analyser and polariser. The graph of the result is shown below.

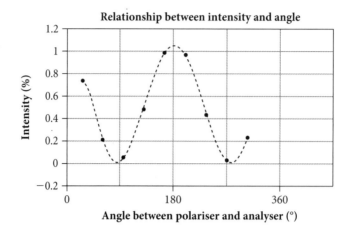

a ⬤◻◻ Using the data in the graph, describe the relationship between the intensity and the angle between the polariser and analyser.   2 marks

_____

_____

_____

_____

b ⬤⬤◻ The student wanted to verify Malus' law by measuring the change in intensity as the angle between the polariser and analyser was changed. What graph would the student plot to verify the relationship from their experiment? Justify your answer.   2 marks

_____

_____

_____

_____

Question 13 (4 marks) ⬤⬤◻

Explain how the light pattern shown in the diagram below is formed.

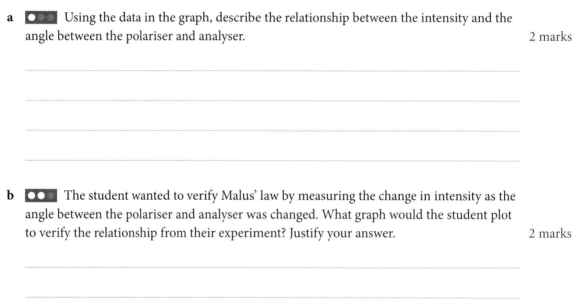

Question 14 (4 marks)

Light from a hydrogen lamp was shone through double slits of separation 2.0 mm, which resulted in the interference pattern shown below.

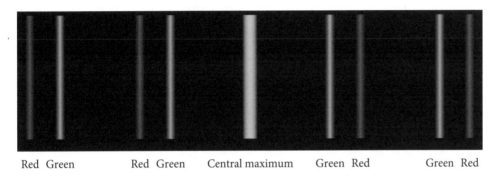

Red Green   Red Green   Central maximum   Green Red   Green Red

The hydrogen spectrum usually consists of four lines but, in this experiment, a yellow filter has been used to allow only the red and green lines to be visible.

Determine the distance between the second-order maxima made by the 486 nm and 656 nm sources, when the screen is 50 cm away.

Question 15 (3 marks)

Two polarising filters are placed at 90° to each other so that no light passes through them, as shown below.

A third, smaller filter is placed between them at 45°. Light can be seen to have travelled through the three filters, as shown below.

Explain why you can now see through the polarisers when a third filter is placed between the first two filters.

_____

_____

_____

_____

_____

_____

Question 16 (4 marks) ●●●

Models can be used to explain and predict phenomena. Diffraction can be explained by modelling visible light as a wave. Below is a section of the electromagnetic spectrum with emission lines shown in the visible range.

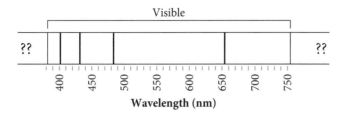

Using diffraction as an example, show that the wave model can be used successfully to make a prediction.

Test 10: Light: quantum model

Section I: 10 marks. Section II: 25 marks. Total marks: 35.
Suggested time: 50 minutes

Section I: Multiple-choice questions

Instructions to students
*   For each question, circle the multiple-choice letter to indicate your answer.

Question 1

A simplified graph of the spectral irradiance of the Sun versus wavelength is shown below.

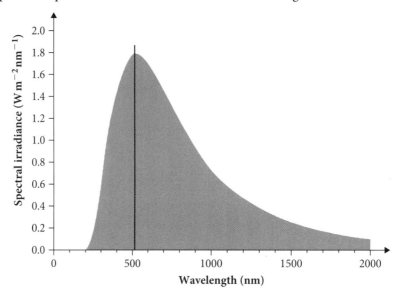

What information is required to determine the surface temperature of the Sun?

A The area under the graph

B The wavelength with the highest spectral irradiance

C The value of the peak spectral irradiance

D The range of wavelengths emitted

Question 2

UV light is shone on a zinc plate that
is connected to an electroscope that
is negatively charged, as shown in
the diagram on the right. The gold
leaves, which were originally
separated, fall together.

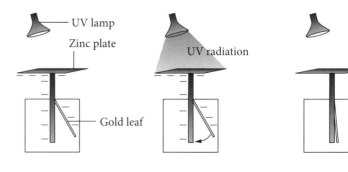

What happens when this experiment
is repeated with red light instead of
UV light?

A The result is the same.

B The leaves don't move.

C The separation increases.

D The separation decreases.

Question 3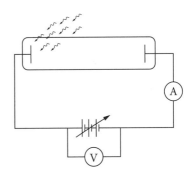

In a black body, when examining the energy emitted, the intensity, as well as the wavelength with the greatest spectral irradiance, are related. Which of the following correctly describes the relationship?

	Temperature	Intensity	Wavelength
A	Increases	Decreases	Decreases
B	Increases	Increases	Increases
C	Decreases	Decreases	Increases
D	Decreases	Increases	Increases

Question 4

Determine the threshold frequency for beryllium.

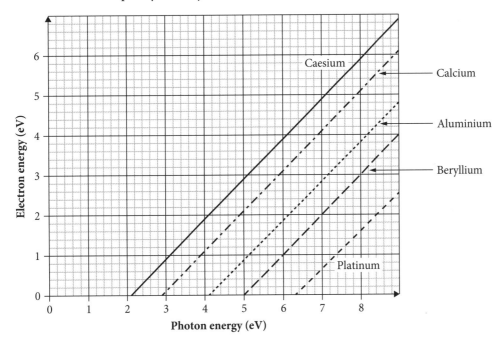

A 5.0 Hz

B 2.1 Hz

C 7.5×10^{34} Hz

D 12×10^{14} Hz

Question 5

Green light is shone on a photoelectric cell. A small current is detected, and the stopping voltage is determined.

The experiment is repeated with violet light that has half the intensity of the green light. What happens to the resulting current and voltage?

	Current	Voltage
A	Increases	Increases
B	Increases	Decreases
C	Decreases	Increases
D	Decreases	Decreases

Use the following information to answer Questions 6 and 7.

The graph below shows the relationship between the frequency of incident light and the kinetic energy of electrons ejected from a variety of metals.

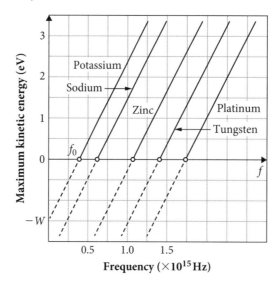

Question 6 ○◐◑

Determine the minimum wavelength of light that would produce photoelectrons for potassium.

A 3.75 m

B 800 nm

C 200 nm

D 3.75×10^{15} m

Question 7 ○◐◑

The difference in work function between sodium and tungsten is closest to

A 4×10^{-19} J

B 9×10^{-19} J

C 5×10^{-19} J

D 0.8 J

Question 8 ©NESA 2013 SIA Q20 ●●●

The graph shows the maximum kinetic energy (E) with which photoelectrons are emitted as a function of frequency (f) for two different metals X and Y.

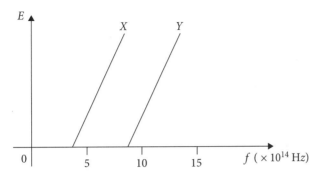

The metals are illuminated with light of wavelength 450 nm.

What would be the effect of doubling the intensity of this light without changing the wavelength?

A For metal X, the number of photoelectrons emitted would not change, but the maximum kinetic energy would increase.

B For metal X, the number of photoelectrons emitted would increase, but the maximum kinetic energy would remain unchanged.

C For both metals X and Y, the number of photoelectrons emitted would not change, but the maximum kinetic energy would increase.

D For both metals X and Y, the number of photoelectrons emitted would increase, but the maximum kinetic energy would remain unchanged.

Question 9 ©NESA 2014 Q11 (ADAPTED) ●●●

Why is there a low intensity of black body radiation at very short wavelengths?

A The energy of each photon is reduced at very short wavelengths.

B There are fewer photons with high energy (i.e. short wavelengths).

C Only photons of very short wavelengths are re-absorbed by the black body.

D Photons of very short wavelengths interact with each other, causing destructive interference.

Question 10 ●●●

Blue light is used to produce a photocurrent, as shown in the diagram. The frequency of the incident light is decreased but its intensity is increased.

What will happen to the readings of the voltage and current?

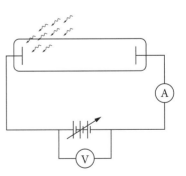

A Voltage will increase but current will decrease.

B Voltage will decrease and current will decrease.

C Voltage will increase and current will increase.

D Voltage will decrease but current will increase.

Section II: Short-answer questions

> **Instructions to students**
> • Answer all questions in the spaces provided.

Question 11 (6 marks) ●▨▨▨

Light is shone on a phototube, as shown in the diagram.

The frequency is increased, and the voltage and current are monitored.

The experiment is repeated, but this time the frequency is set to the lowest frequency at which current is detected. The results are shown in the graphs below.

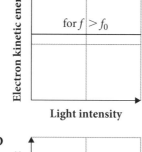

A
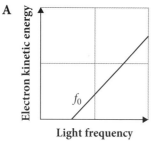

B
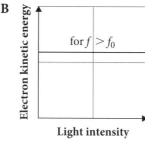

C
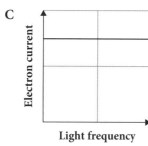

D
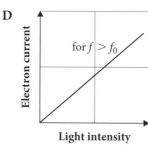

How do these results support the idea of the quantum nature of light?

_____

_____

_____

_____

_____

_____

_____

_____

_____

_____

Question 12 (3 marks)

How does Planck's mathematical model of the black body radiation curve provide evidence for the quantum nature of light?

_____

_____

_____

_____

_____

_____

Question 13 (7 marks)

A simulator was used to determine the relationship between temperature and the peak wavelength of light emitted by a black body. The table of results is below.

Intensity (W m^{-2})	Peak wavelength (μm)	Temperature (K)
4.59×10^6	0.966	3000
2.33×10^7	0.644	5000
7.11×10^7	0.487	5950
2.11×10^8	0.367	7900
5.56×10^8	0.291	9950

a On the grid provided, draw a graph showing the relationship between temperature and peak wavelength.   3 marks

b Using your graph, determine the temperature if the peak wavelength is 0.8 μm.   2 marks

_____

_____

c   Briefly discuss how the other piece of evidence in the data provided supports the model
of the black body curve.   2 marks

_____

_____

_____

_____

Question 14 (4 marks) [○○■]

Light of wavelength 599 nm is shone on copper that has a work function of 4.53 eV. Determine
whether an electron will be ejected from the surface of the copper.

_____

_____

_____

_____

_____

_____

Question 15 (5 marks)

The temperature of a black body can be determined from the peak wavelength with peak
spectral irradiance.

a   [○○■]   Using the axes below, sketch three curves that show how the black body curve
changes with temperature.   3 marks

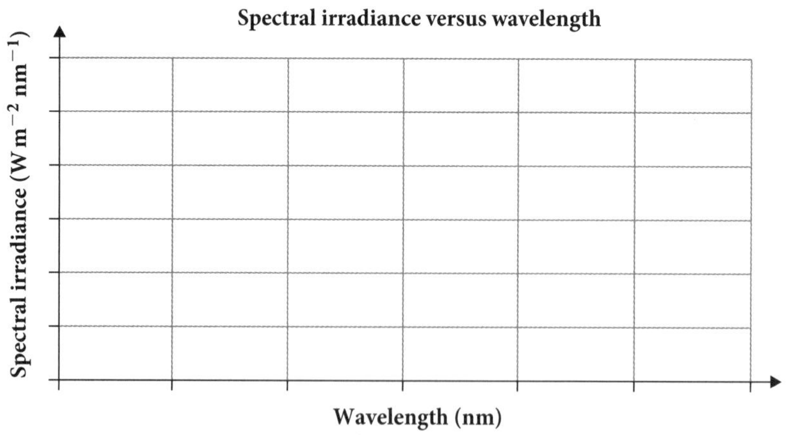

Spectral irradiance versus wavelength

y-axis: Spectral irradiance (W m^{-2} nm^{-1})

x-axis: Wavelength (nm)

b   [○○■]   Outline the method used to determine the temperature from one of these curves.   2 marks

_____

_____

Test 11: Light and special relativity

Section I: 10 marks. Section II: 25 marks. Total marks: 35.
Suggested time: 50 minutes

Section I: Multiple-choice questions

Instructions to students
- For each question, circle the multiple-choice letter to indicate your answer.

Question 1

How much energy is released if 5.0 kg of mass is converted into energy?

A 1.5×10^9 J

B 4.5×10^{17} J

C 1.5×10^{12} J

D 4.5×10^{15} J

Question 2

You are sleeping on a train that is travelling at constant velocity and you wake up in a dark tunnel. To determine if you are still moving, you could

A hold up an accelerometer and see if it deviated from the centre.

B drop a ball and see if it fell behind you.

C hold up a mirror to see if you still had a reflection.

D do nothing because there is no way to tell from within your frame of reference.

Question 3

In Einstein's thought experiment, shown in the diagram below, a person at X is standing at the midpoint of a train carriage that is travelling at $0.9c$. They turn on a light, which triggers a clock at either end of the carriage by way of a sensor. The person sees the clocks start at the same time.

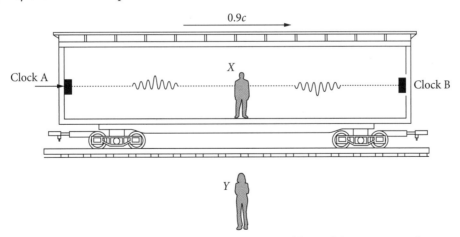

NESA 2020 SI Q17 (ADAPTED)

Determine what a bystander at the train station, position Y, would see if they were standing directly in line with the person on the train when the light was turned on.

A The clocks start at the same time.

B Clock B starts before clock A.

C Clock A starts before clock B.

D They can't see the clocks.

Question 4 ©NESA 2013 SIA Q19 ●●

A spaceship moves close to the speed of light, relative to a planet.

The rest-frame length of the spaceship can be determined by an observer who is

A on the spaceship measuring the time taken for light to travel between two points on the planet.

B on the planet measuring the time taken for light to travel from the front to the back of the spaceship.

C on the spaceship measuring the time taken for light to travel from the front to the back of the spaceship.

D on the planet measuring the difference in the arrival time of light from the front and the back of the spaceship.

Question 5 ●●

Which of the following is **not** an inertial frame of reference?

A A space capsule parachuting at terminal velocity through the atmosphere

B A space station in orbit

C A hypothetical train travelling at $0.5c$

D A lift in a building travelling at a constant rate of change of displacement

Question 6 ●●

Supernovae are giant stellar explosions, the debris from which can move at up to $10\,000\,\mathrm{km\,s^{-1}}$. This debris emits light that can be measured on Earth, and spectroscopic analysis can be used to determine the chemical composition of the debris.

What other statements can be made about the light from this debris as measured from Earth?

A The frequency will be smaller than expected, but the wavelength will be the same.

B The frequency will be larger than expected, but the wavelength will be the same.

C The frequency will be smaller than expected, and the wavelength will be larger than expected.

D The frequency will be larger than expected, but the wavelength will be the shorter than expected.

Question 7 ●●

An electron and a positron, which has the same mass as an electron, collide and annihilate, producing two gamma photons.

What is the frequency of each of the photons?

A $8.14 \times 10^{-14}\,\mathrm{Hz}$

B $1.23 \times 10^{20}\,\mathrm{Hz}$

C $2.47 \times 10^{20}\,\mathrm{Hz}$

D $1.63 \times 10^{-13}\,\mathrm{Hz}$

Question 8 ©NESA 2004 SIA Q6 ●●●

A ball is dropped by a person sitting on a vehicle that is accelerating uniformly to the right, as shown by the arrow.

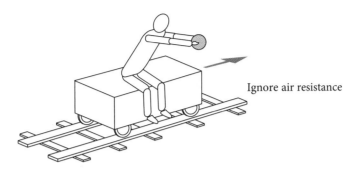

Ignore air resistance

Which of the following represents the path of the ball, shown at equal time intervals, observed from the frame of the reference of the vehicle?

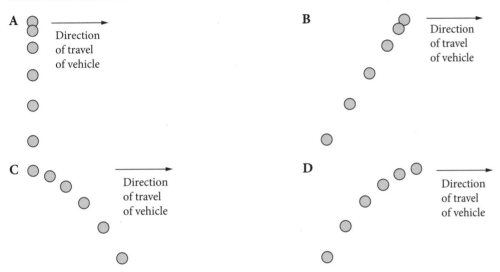

Question 9 ●●

A spaceship travels 1000 km at one-third of the speed of light, and the time it takes is recorded by an external observer.

Which of the following correctly identifies the length and the distance measured by a person on the spacecraft?

	Time (ms)	Length (km)
A	10.6	1060
B	10.6	943
C	9.4	1060
D	9.4	943

Question 10 ©NESA 2020 SI Q17 ● ●

In a thought experiment, observer X is on a train travelling at a constant velocity of $0.95c$ relative to the ground. Observer Y is standing on the ground outside the train. As observer X passes observer Y, observer X sends a short light pulse towards the sensor.

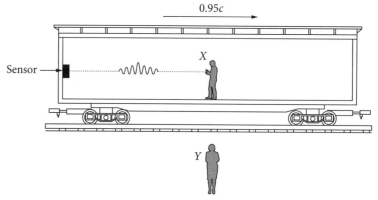

Which statement about the light pulse is correct as observed by X or Y in their respective frames of reference?

A Its velocity observed by Y is $0.05c$.

B X sees it travel a shorter distance to the sensor than Y.

C X sees it take a longer time to reach the sensor than Y.

D Both X and Y see it travel the same distance in the same amount of time.

Section II: Short-answer questions

> **Instructions to students**
> •   Answer all questions in the spaces provided.

Question 11 (5 marks)

The Sun has a power output of 3.8×10^{26} W, which is from the fusion of hydrogen to helium. The process is called the proton–proton chain and can be summarised as

$$4{}^{1}\text{H}^{+} + 2\text{e}^{-} \rightarrow {}^{4}\text{He}^{2+} + 2\nu_{\text{e}}$$

The amount of energy released in this reaction is approximately 4.0×10^{-12} J.

a ● ● ● Determine the loss in mass in this reaction.   1 mark

b ● ● ● Determine the mass lost in 1.0 s by the Sun.   2 marks

c   One kilogram of petrol will combust to produce 46 MJ of energy. How much matter is lost in fusion in the Sun for the same amount of energy produced?   2 marks

Question 12 (3 marks)

An astronaut in the International Space Station, travelling at $7.5\,km\,s^{-1}$, wanted to compare their measurement for the time taken to pass over Singapore to Sydney with a ground-based measurement.

The distance between Singapore and Sydney is 6300 km. A ground-based observer measures the time as $\dfrac{6300}{7.5} = 840\,s$.

a   What is this distance as measured by the astronaut?   2 marks

b   How long would this take, as measured by the astronaut?   1 mark

Question 13 (4 marks)

In the movie *Angels and Demons,* some CERN scientists have a canister with 0.25 g of stolen antimatter. The claim is that this can produce the same amount of energy as 5000 tonnes of dynamite.

If 1 tonne of dynamite equates to 4.184 GJ of energy, assess the validity of this claim.

Question 14 (5 marks)

Muons are created in the upper atmosphere 10 km above the surface of Earth and travel down to Earth at 0.98c. With a half-life of 1.56 μs, it would be predicted that a very low number of them – only 0.3 per million – would be detected in laboratories at sea level. However, considerably more are actually detected than is predicted: over 49 000 per million.

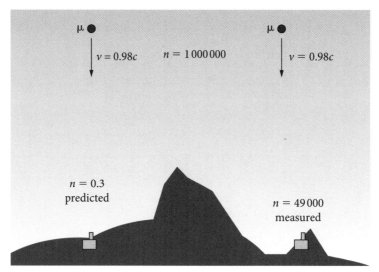

a   Given that a muon's half-life measured in a laboratory is 1.56×10^{-6} s, how far would muons be able to travel at 0.98c in this time?   1 mark

b ⬤⬤ Referring to your answer to part **a**, explain why only 0.3 muons per million might be expected to reach Earth's surface.   2 marks

c ⬤⬤ Explain why more muons are measured than might be expected.   2 marks

Question 15 (4 marks) ⬤⬤

The graph below shows the momentum of an object versus its speed.

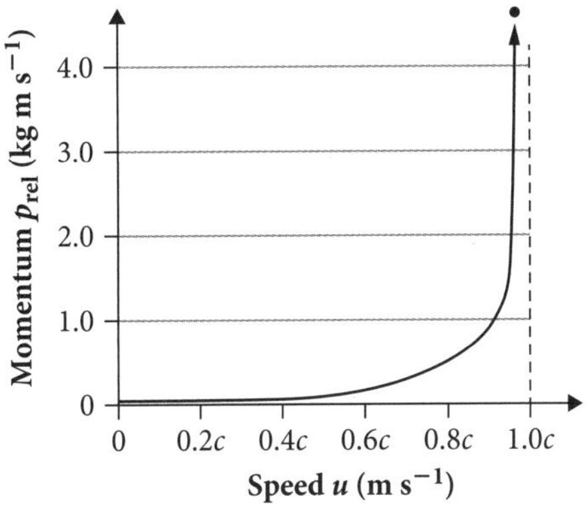

a Using the graph and equations, describe the relationship between the momentum of an object and its speed.   2 marks

b Why is the relativistic formula not used to calculate the momentum in a car crash?   2 marks

Question 16 (4 marks) ⬤⬤⬤

Explain how experimental data can validate models in reference to Einstein's special theory of relativity.

CHAPTER 4
MODULE 8: FROM THE UNIVERSE TO THE ATOM

Test 12: Origins of the elements

Section I: 10 marks. Section II: 25 marks. Total marks: 35.
Suggested time: 60 minutes

Section I: Multiple-choice questions

Instructions to students
- For each question, circle the multiple-choice letter to indicate your answer.

Question 1

A fusion reaction is shown below.

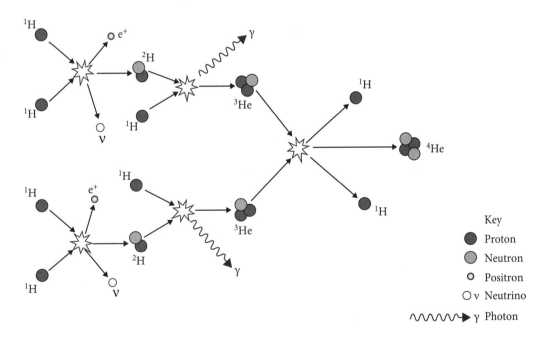

The net result of this reaction is best represented as

A $6\,^1H \rightarrow He$

B $^1H \rightarrow 4\,^4He$

C $4\,^1H \rightarrow\,^4He$

D $2\,^3He \rightarrow\,^4He$

Question 2 ©NESA 2019 SI Q4

Four stars, *P*, *Q*, *R* and *S*, are labelled on the Hertzsprung–Russell diagram below.

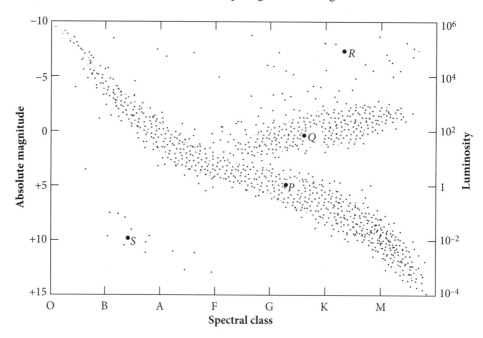

Which statement is correct?

A *S* has a greater luminosity than *Q*.

B *R* is a blue star whereas *S* is a red star.

C *S* has a higher surface temperature than *R*.

D *P* is at a more advanced stage of its evolution than *R*.

Use the Hertzsprung–Russell diagram of stars below to answer Questions 3–5.

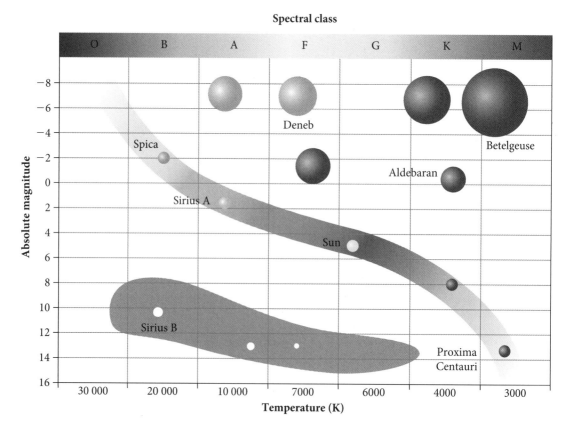

Question 3

A large star is found to have a spectral class of B. What are the characteristics of this star?

A A blue star that has high luminosity

B A hot white dwarf that has low luminosity

C A main sequence star that has low luminosity

D A cool star that has high luminosity

Question 4

Which statement is true about the H–R diagram?

A A larger absolute magnitude means a brighter star.

B As stars age, they move to the left of the diagram.

C A random selection of stars found on the main sequence could all have the same age.

D Stars are always older if they are found to the right of the diagram.

Question 5

Spica and the Sun are both main sequence stars and therefore both fuse hydrogen to produce helium. Spica has around 20 times greater luminosity than the Sun and is significantly hotter. The best explanation for this is

A Spica is denser than the Sun.

B Spica is larger than the Sun.

C Spica is more blue than the Sun.

D Spica is closer than the Sun.

Question 6

A graph that Hubble used to determine the age of the Universe is shown below.

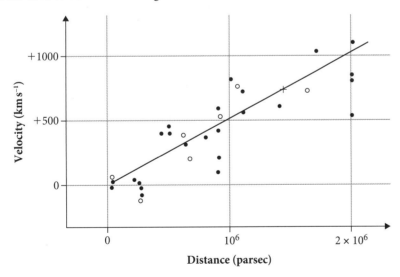

Given that 1 parsec = 3.086×10^{13} km, what is the age of the Universe using Hubble's data?

A 1.9 billion years

B 13.7 billion years

C 6.17 billion years

D 71 billion years

9780170465298

Use the Hertzsprung–Russell diagram below to answer Questions 7 and 8.

Schematic Hertzsprung–Russell diagram

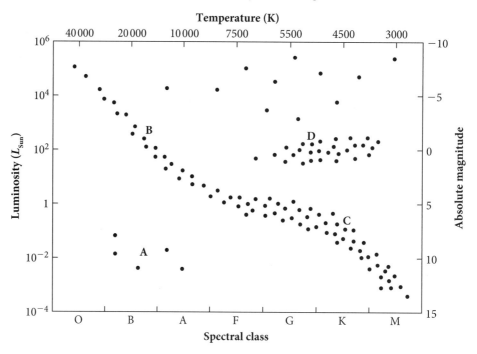

Question 7 ⬤⬤

Where would a white dwarf be best positioned on an H–R diagram?

A A

B B

C C

D D

Question 8 ⬤⬤

If a star has the same spectral class as the Sun but is 100 times more luminous, where would it be placed on the H–R diagram?

A A

B B

C C

D D

Question 9 ⬤⬤◼

The diagram below shows the spectra taken from two galaxies, NGC 3147 and NGC 3368.

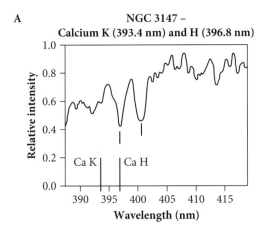

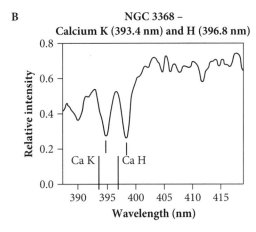

The wavelengths of the spectra from calcium K and calcium H are determined in the laboratory and are shown by lines at the bottom of each graph. They do not match the spectra from the galaxies. What can be concluded from this data about the galaxies?

A The galaxies are moving away from us.

B The galaxies are moving towards us.

C The peaks represent other elements.

D Calcium is not present in these galaxies.

Question 10 ⬤⬤⬤

According to Georges Lemaître's Big Bang theory, radiation that was expanding eventually became matter, as seen in the graph.

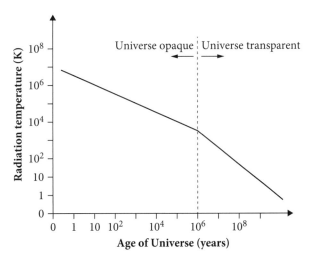

What occurred 1 million years after the Big Bang?

A A sudden drop in the density of the Universe, referred to as the inflationary period

B The formation of stars and galaxies

C The production of matter from energy

D A sudden drop in temperature that caused a decrease in the density of the Universe

Section II: Short-answer questions

Instructions to students
- Answer all questions in the spaces provided.

Question 11 (5 marks)

How can the following data be interpreted as evidence for the expansion of the Universe?

CMB image

Hubble's velocity–time relationship

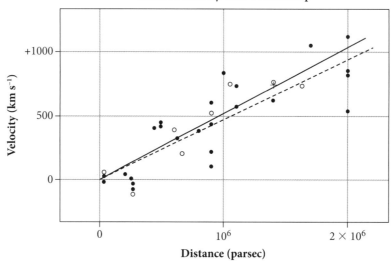

The ten most abundant elements in the Universe

Element	Abundance measured relative to silicon
Hydrogen	40 000
Helium	3100
Oxygen	22
Neon	8.6
Nitrogen	6.6
Carbon	3.5
Silicon	1
Magnesium	0.91
Iron	0.6
Sulfur	0.38

© ESA and the Planck Collaboration

Source : Exploring Chemical Elements and
their Compounds David I. Heiserman, 1992

_____

_____

_____

_____

_____

_____

_____

_____

_____

Question 12 (5 marks)

Main sequence stars and red giants both derive their energy from fusion, but the main type of fusion reaction is different for each.

Compare a G2 Main Sequence star with a K1 red giant in relation to these two processes. Relate this to their characteristics.

Fusion reaction	Energy released per reaction (MeV)
p–p chain	1.44
Triple alpha	7.2

_____

_____

_____

_____

_____

_____

_____

_____

Question 13 (3 marks) ●●◖

The spectrum of the Sun is shown below.

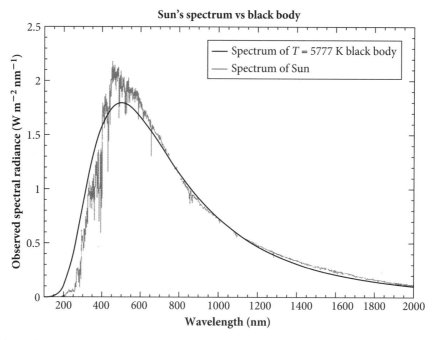

Explain how this spectrum is produced by the Sun.

_____

_____

_____

_____

_____

Question 14 (4 marks) ●● ▪

A tweet by Professor Geraint Lewis of the University of Sydney is shown.

Geraint Lewis

Assess how well this models the Big Bang theory.

_____

_____

_____

_____

_____

_____

_____

_____

_____

_____

Question 15 (5 marks) ⬤⬤⬤

CNO cycle and p–p chain are fusion reactions that occur in main sequence stars.

A graph incorporating the luminosity of stars versus temperature is below. The core temperature of the Sun is 15 million K.

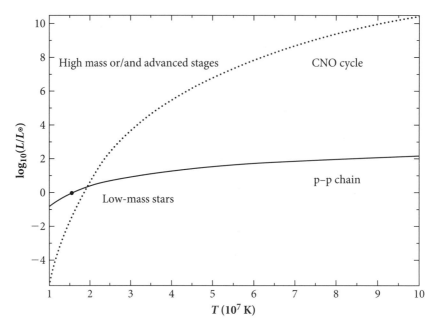

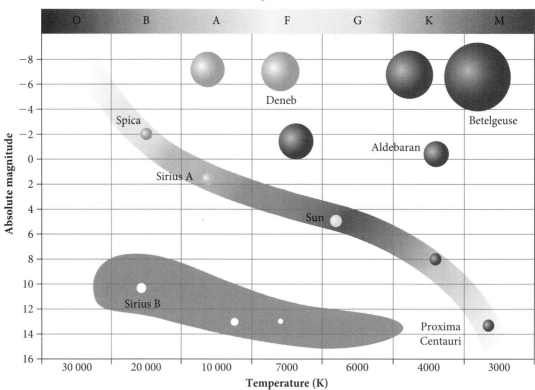

Using the information provided, explain how the dominance of each fusion reaction may vary in stars along the main sequence. Cite two examples in your response.

_____

_____

_____

_____

_____

_____

_____

_____

_____

_____

Question 16 (3 marks)

Explain, with reference to a specific stage of stellar evolution, why a star is stable in size and brightness.

_____

_____

_____

_____

_____

_____

Test 13: Structure of the atom

Section I: 10 marks. Section II: 25 marks. Total marks: 35.
Suggested time: 60 minutes

Section I: Multiple-choice questions

Instructions to students
- For each question, circle the multiple-choice letter to indicate your answer.

Question 1 ©NESA 2019 SI Q3 ○●●

Geiger and Marsden carried out an experiment to investigate the structure of the atom.
Which diagram identifies the particles they used and the result that they **initially** expected?

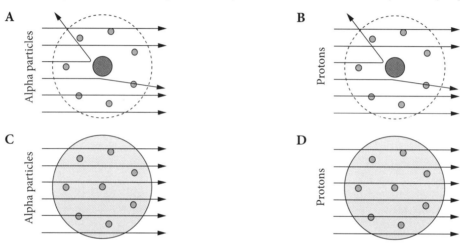

Question 2 ○●●

Early experiments to determine the nature of cathode rays helped identify some characteristics of electrons. One such experiment is shown below.

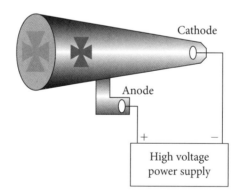

The characteristic determined from this was that cathode rays

A were negatively charged.

B travel in straight lines.

C were travelling as waves.

D did not contain electrons.

Question 3 ⬤◽◽

Which of the following experiments with cathode rays helped determine the negative nature of the electron?

I    II 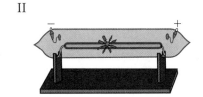   III

A I only

B I and II

C I, II and III

D II and III

Question 4 ⬤⬤◽

Which one of the following correctly describes the experimental set-up of Robert Millikan?

A Suspending a positively charged particle with an electric field going upwards

B Suspending a positively charged particle with an electric field going downwards

C Suspending a negatively charged particle with an electric field going upwards

D Suspending a negatively charged particle with an electric field going downwards

Question 5 ⬤⬤◽

Rutherford's model of the atom was preferred over Thomson's model. Which of the following are reasons for this?

A Rutherford's model could explain spectral lines.

B Rutherford's model could explain why alpha particles could pass though gold.

C Rutherford's model could explain the recoil of alpha particles.

D The structure of the atom is consistent with a planetary model.

Question 6 ⬤⬤◽

Thomson's experiment contributed to our understanding of the atom. Which of the following best describes his contribution? He determined the

A mass and charge of the electron.

B charge on the electron by knowing the density of oil drops.

C mass of the electron from the density of oil drops.

D charge-to-mass ratio of the electron.

Question 7

In Chadwick's experiment, a strange radiation was emitted when a boron target was bombarded with alpha particles.

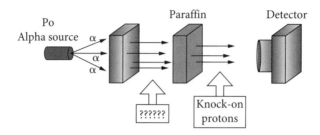

What characteristics of the new radiation did not fit with any known radiation?

A Its mass was larger than alpha particles.

B It could not be deflected by magnetic fields and yet had mass.

C It was negatively charged.

D It had no momentum.

Question 8

Millikan's oil-drop experiment used electric fields to help determine the charge of the electron. When the oil drop was in the electric field, it moved at a constant velocity.

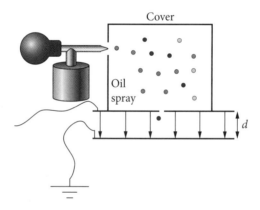

Which expression allows the characteristics of an electron to be determined from the measurements that could be made with this apparatus?

A $q = \dfrac{mg}{Vd}$

B $q = \dfrac{d}{Vmg}$

C $q = \dfrac{Vd}{mg}$

D $q = \dfrac{mgd}{V}$

Use the diagram below to answer Questions 9 and 10.

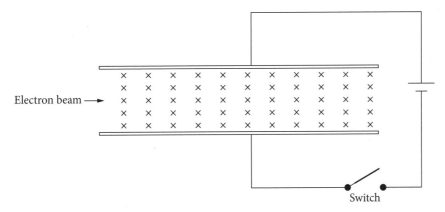

Question 9 ⬤⬤

With the switch open, given the charge-to-mass ratio is $1.76 \times 10^{11}\,\text{C kg}^{-1}$, and the velocity of the beam is $2.0 \times 10^{7}\,\text{m s}^{-1}$, what is the magnetic field strength needed to cause a deflection with a radius of 3.0 cm?

A $3.4\,\mu\text{T}$

B $38\,\mu\text{T}$

C $3.4 \times 10^{-4}\,\text{T}$

D $3.8\,\text{mT}$

Question 10 ⬤⬤⬤

The switch is now closed and the beam curves upwards. Which of the following is most likely to be true?

A The force from the magnetic field is stronger than the force from the electric field and thus the curve is circular.

B The force from the electric field is stronger than the force from the magnetic field and thus the curve is parabolic.

C The force from the magnetic field is stronger than the force from the electric field but is neither parabolic nor circular.

D The force from the electric field is stronger than the force from the magnetic field but is neither parabolic nor circular.

Section II: Short-answer questions

Instructions to students
- Answer all questions in the spaces provided.

Question 11 (3 marks)

Thomson's model of the atom was used until evidence suggested it needed to be refined.

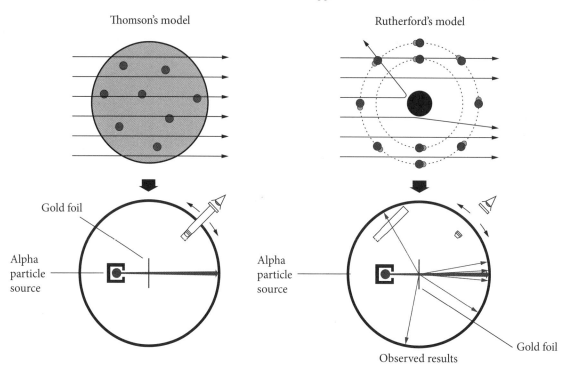

Outline how the results from the Geiger–Marsden experiment led to the Rutherford model of the atom.

_____

_____

_____

_____

_____

Question 12 (3 marks)

The Millikan oil-drop experiment was used to measure the charge on an electron. Below is an image of a suspended oil drop and data from an oil-drop experiment. Use the data given in the table to determine the charge.

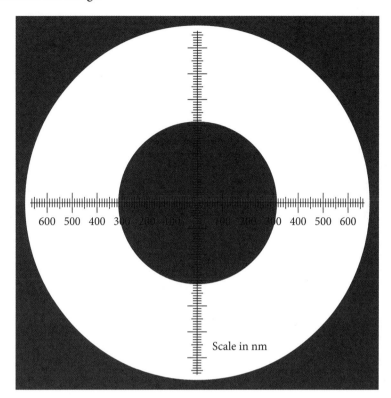

| 600 | 500 | 400 | 300 | 200 | 100 | | 100 | 200 | 300 | 400 | 500 | 600 |

Scale in nm

Voltage between plates	11.7 V
Plate separation	3.0 mm
Density of the oil drop	900 kg m^{-3}
Volume of a sphere	$V = \frac{4}{3}\pi r^3$

Question 13 (6 marks) ©NESA 2014 SIB Q28 (ADAPTED) ⬤⬤

a   Thomson's experiment measures the charge/mass ratio of an electron.

Use an annotated diagram to show the experimental setup for Thomson's experiment to be performed.   3 marks

b   An electron is projected at 90° into a magnetic field of 9×10^{-4} T, at a speed of 1×10^7 m s^{-1}. This causes the electron to undergo uniform circular motion.

Calculate the radius of the electron's path.   3 marks

Question 14 (4 marks)

Students were asked to undertake an activity that simulated the Geiger–Marsden experiment. They rolled a ball under a piece of cardboard that already had an object of unknown shape underneath it, as shown in the diagram below.

They recorded where the ball emerged.

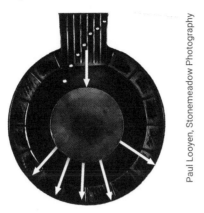

Paul Looyen, Stonemeadow Photography

Their results are shown in the diagram below.

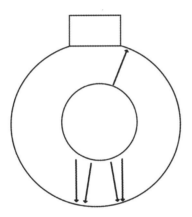

Discuss whether this is an effective representation of the Geiger–Marsden experiment.

_____

_____

_____

_____

_____

_____

_____

_____

Question 15 (2 marks) ◖◖◗

In 1911 Rutherford discovered the nucleus, and in 1919 he determined that the nucleus of a hydrogen atom was a discrete particle, later called a proton. He therefore deduced that the nuclei of atoms were made up of protons.

However, his model left some things unexplained. For example, helium was known to have an atomic number of 2 but a mass number of 4.

Analyse this information with reference to a suggested new subatomic particle.

Question 16 (7 marks) ◖◖◗

The scientific method involves the building of models based on observations. However, these models are subsequently tested and if new observations are not consistent with a model, it needs to be revised or rejected.

With reference to two scientists' work, show how this was applied to the development of the understanding of the atom.

Test 14: Quantum mechanical nature of the atom

Section I: 10 marks. Section II: 25 marks. Total marks: 35.
Suggested time: 50 minutes

Section I: Multiple-choice questions

Instructions to students
- For each question, circle the multiple-choice letter to indicate your answer.

Question 1

Which statement is correct regarding the Bohr model of the atom?

A It explained the reason for stable orbits.

B It provided an explanation for emission spectra.

C It was the first fully quantum model of the atom.

D It formed the basis for the Rutherford model.

Use the following information to answer Questions 2 and 3.

The diagram below shows the Balmer series of the hydrogen emission spectrum.

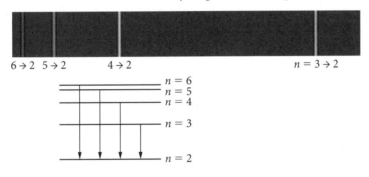

Question 2

The wavelength of the red line (the line on the far right) in the spectrum is calculated to be

A $1.82 \times 10^{6}\,\text{m}$

B $1.52 \times 10^{6}\,\text{m}$

C $5.47 \times 10^{-7}\,\text{m}$

D $6.56 \times 10^{-7}\,\text{m}$

Question 3

Determine the energy needed for the electron to produce the violet line (the first line on the spectrum).

A 7.31×10^{14} J

B 2.7×10^{-40} J

C 4.84×10^{-19} J

D 7.2×10^{-19} J

Question 4 ▮▮

De Broglie suggested that electrons in their orbits around hydrogen can be considered as standing waves. The diagram shows the representation of the matter wave for $n = 2$, where the electron is not in its ground state.

If the radius in a hydrogen atom is 9.99×10^{-10} m, calculate the wavelength of the matter wave.

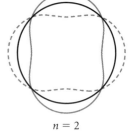

$n = 2$

A 0.999 nm

B 6.28 nm

C 12.6 nm

D 3.14 nm

Question 5 ▮▮

An electron in a cathode ray has a velocity of $5.0 \times 10^7 \, \text{m s}^{-1}$. Calculate the wavelength of the electron.

A 7.2×10^{-4} m

B 1.5×10^{-11} m

C 6.8×10^{10} m

D More information is required.

Question 6 ©NESA 2020 SI Q9 ▮▮

Bohr improved on Rutherford's model of the atom.

Which observation by Bohr provided evidence supporting his improvement?

A Elements produced unique emission spectra consisting of discrete wavelengths.

B The collision of an electron and a positron produced two photons that travelled in opposite directions.

C A small percentage of alpha particles fired at a gold foil target were deflected by angles of more than 90 degrees.

D A beam of electrons reflected from a nickel crystal produced a pattern of intensity at different angles, consistent with their wave properties.

Question 7

Type	Name	Wavelength (nm)
Ultraviolet	UV-C	100–280
	UV-B	280–315
	UV-A	315–400
Visible light	Purple	400–435
	Blue	435–480
	Patina	480–490
	Blue-green	490–500
	Green	500–560
	Yellow-green	560–580
	Yellow	580–595
	Orange	595–610
	Red	610–750
	Red-purple	750–800
Infrared	IR-A	800–1400
	IR-B	1400–3000
	IR-C	$3000–10^6$

Name the electromagnetic wave that is emitted when an electron moves from $n = 4$ to $n = 3$.

A UV-C

B UV-A

C IR-B

D IR-A

Question 8

Schrödinger's model of the atom is more comprehensive than previous models of the atom. Which of the following statements best describes an important aspect of his model?

A The position of each electron is described as a probability.

B The nucleus is a central small mass.

C Electrons orbit the nucleus in distinct energy levels.

D The atom is a positive mass with removable negative charges.

Question 9

Schrödinger's model of the atom developed from Bohr's model. Images of their models are shown on the right.

The electrons in each model can be best characterised as being located in

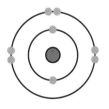

	Bohr	Schrödinger
A	Definite shells	Regions of energy
B	Energy levels	Bands of energy
C	Energy orbitals	Probability regions
D	Probability regions	Energy levels

Question 10 ⬤⬤⬤

Bohr's explanation of spectral lines is consistent with the conservation of energy because the energy of the light

A absorbed by the atom gradually increases the energy of electrons until they are in orbits consistent with the Rydberg equation.

B released by the atom causes a specific drop in electron energy levels.

C absorbed or released by the atom is limited to specific values consistent with the Rydberg equation.

D released by the atom must be less than the specific values absorbed, consistent with the Rydberg equation.

Section II: Short-answer questions

Instructions to students
• Answer all questions in the spaces provided.

Question 11 (5 marks) ⬤⬤⬤

a Calculate the minimum energy that can be absorbed by a hydrogen atom at energy level 2.   2 marks

b Explain why the emission spectra observed by Balmer provided evidence for the quantum nature of the atom.

3 marks

Question 12 (2 marks) ⬤⬤⬤

De Broglie suggested that electrons can behave as matter waves. Determine the velocity at which an electron would be moving if its wavelength was 6.56×10^{-7} m.

Question 13 (5 marks) ⬤⬤◯

Assess the limitations of the Rutherford and Bohr models of the atom.

_____

_____

_____

_____

_____

_____

_____

_____

_____

_____

Question 14 (3 marks) ⬤⬤◯

The image below shows the results obtained by Davisson and Germer's electron-diffraction experiment.

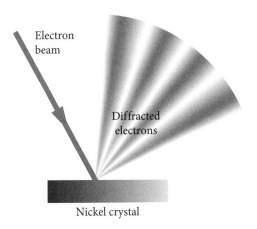

How do the results support de Broglie's proposal of a wave nature for electrons, and how did this contribute to the model of the atom?

_____

_____

_____

_____

_____

_____

Question 15 (4 marks) `●●`

Schrödinger's contribution to the current model of the atom is more comprehensive than Bohr's model, yet we are still taught Bohr's model when learning about the atom. Explain why we use models that are not fully accurate to understand phenomena.

Question 16 (6 marks) `©NESA 2013 SII Q35e` `●●●`

How did de Broglie use existing concepts and ideas to come up with new interpretations that have increased our understanding of the structure of matter?

Test 15: **Properties of the nucleus**

Section I: 10 marks. Section II: 25 marks. Total marks: 35.
Suggested time: 60 minutes

Section I: Multiple-choice questions

Instructions to students
- For each question, circle the multiple-choice letter to indicate your answer.

Question 1 ©NESA 2020 SI Q4 ⬤▨▨

The graph shows the mass of a radioactive isotope as a function of time.

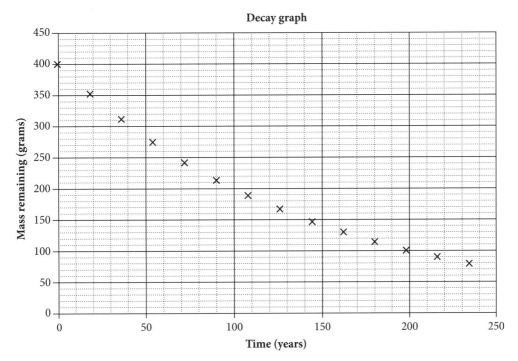

What is the decay constant, in years^{-1}, for this isotope?

A 0.0030

B 0.0069

C 2.0

D 100

Question 2 ⬤▨▨

Three particles enter a magnetic field as shown in the diagram on the right.

Which of the following correctly identifies the particles?

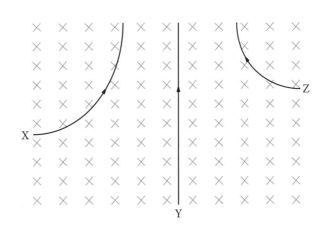

	X	**Y**	**Z**
A	Alpha	Beta	Gamma
B	Alpha	Gamma	Beta
C	Beta	Gamma	Alpha
D	Beta	Alpha	Gamma

Question 3 ⬤▨▨

Which statement best describes nuclear fission, in which energy is released?

A It can occur with any nuclei as it splits into smaller atoms.

B There is a loss of nucleon(s) in fission according to $E = mc^2$.

C The binding energy per nucleon is greater for each of the products than the reactants.

D The combined mass of the products is greater than reactants.

Question 4 ⬤▨▨

Which of the following components in a nuclear fission reactor is responsible for absorbing neutrons?

A Moderator

B Control rods

C Heat exchanger

D Core

Question 5   ©NESA   2020 SI Q8   ⬤⬤

A uranium isotope, U, undergoes four successive decays to produce Q.

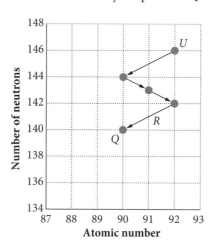

Which row of the table correctly shows the decay process R and product Q?

	Process R	**Product Q**
A	α	Pa-230
B	β	Pa-234
C	α	Th-230
D	β	Th-234

Question 6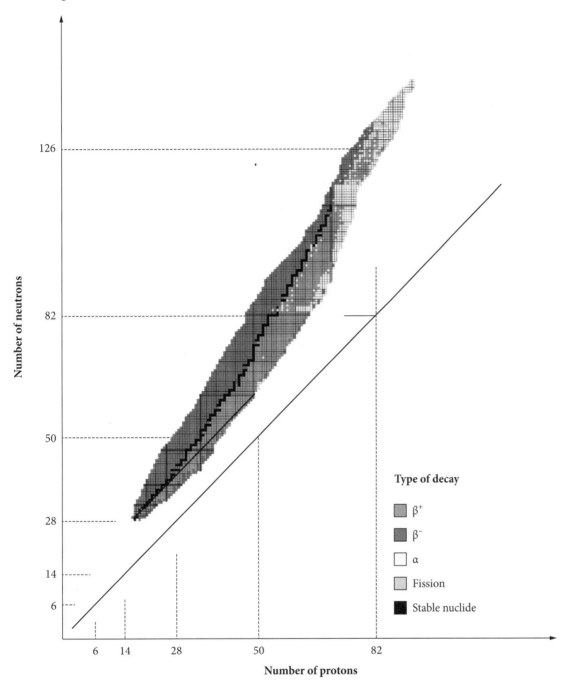

The graph below plots the number of neutrons versus the number of protons for elements in the periodic table. It shows which atoms are stable and how unstable atoms will decay to become stable. The diagonal black line represents $N = Z$.

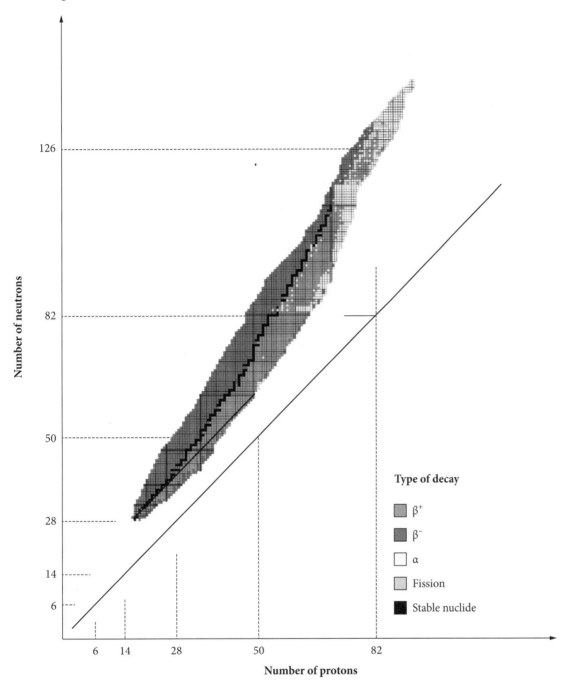

Nickel-56 is an unstable isotope. To become more stable, it will undergo transmutation and form a new element.

Identify the correct transmutation and the element it will become.

A Beta, copper

B Beta, cobalt

C Alpha, copper

D Alpha, cobalt

Question 7 ●○○

Radioactive sources can be used to detect leaks in pipes that are underground. Radioactive sources are emitters of radiation that are injected into the pipes. A Geiger counter is used to detect the radiation that is released from the source, as shown in the diagram below.

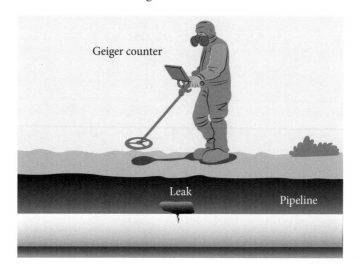

The emitter that is used would most likely be

A a beta source because beta radiation can penetrate deep ground and is short lived.

B an alpha source because alpha radiation is easier to detect.

C a gamma source because gamma radiation can penetrate deeply though most things.

D any radioactive source because a Geiger counter can detect anything.

Question 8 ○●○

A sample of ^{226}Ra takes 1622 years to reduce to half its original amount. If there is 35 g of radium in the initial sample, how much would remain after 3000 years?

A 7.9 g

B 18.9 g

C 9.7 g

D 20.4 g

Use the following information to answer Questions 9 and 10.

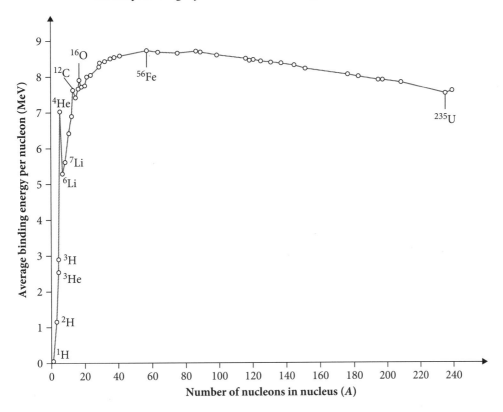

Question 9 ⬤⬤▢

Which of the following simplified reactions will release the most energy?

A H + H → He

B H + Li → Be

C U → Ba + Kr

D Fe + Fe → Te

Question 10 ⬤⬤▢

Calcium will only be a 'useful' energy source if fusion (rather than fission) becomes a method of energy production. This is because its binding energy

A increases when fused and therefore releases energy.

B decreases when fused and therefore releases energy.

C increases when fused and therefore absorbs energy.

D decreases when fused and therefore absorbs energy.

Section II: Short-answer questions

> **Instructions to students**
> • Answer all questions in the spaces provided.

Question 11 (3 marks)

A graph of the percentage of ^{60}Co remaining after a period of time is shown below.

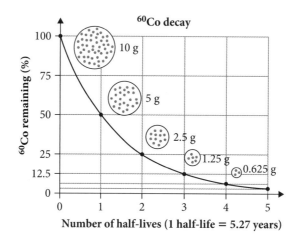

a ◼◻◻◻ How much time has passed if 1.25 g of ^{60}Co remains?   1 mark

b ◼◼◻◻ Determine the percentage of atoms remaining after 14.82 years.   2 marks

Question 12 (3 marks)

Radon-222 spontaneously undergoes alpha decay.

a ◼◻◻◻ What is the nuclear equation for this reaction?   2 marks

b ◼◼◻◻ The mass of the products is less than the mass of the reactants. The resulting mass defect leads to the release of energy. What happens to this energy?   1 mark

Question 13 (3 marks)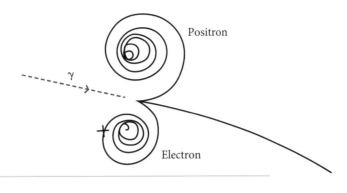

A gamma photon strikes a hydrogen nucleus in a cloud chamber that is experiencing a magnetic field into the page. The result is the production of two beta particles: a positron and an electron.

Explain what can be learned from the traces.

_____

_____

_____

_____

_____

_____

Question 14 (4 marks)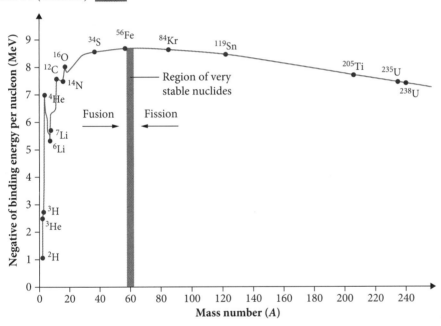

Explain why some nuclides release energy by undergoing fusion, whereas others only release energy by undergoing fission.

_____

_____

_____

_____

_____

_____

_____

Question 15 (5 marks) ⬤⬤◑

A person living near a nuclear reactor is heard to say 'I'm concerned that if there is an accident, we will have a nuclear bomb on our doorstep'.

Using physics principles, assess this statement.

Question 16 (7 marks) ©NESA 2019 SII Q36 ⬤⬤⬤

A radon-198 atom, initially at rest, undergoes alpha decay. The masses of the atoms involved are shown in atomic mass units (u).

$$\text{radon-198} \quad \rightarrow \quad \text{polonium-194} \quad + \quad \text{helium-4}$$
$$197.999\ u \qquad\qquad 193.988\ u \qquad\qquad 4.00260\ u$$

The kinetic energy of the polonium atom produced is 2.55×10^{-14} J.

By considering the mass defect, calculate the kinetic energy of the alpha particle, and explain why it is significantly greater than that of the polonium atom.

Test 16: Deep inside the atom

Section I: 5 marks. Section II: 20 marks. Total marks: 25.
Suggested time: 40 minutes

Section I: Multiple-choice questions

Instructions to students
- For each question, circle the multiple-choice letter to indicate your answer.

Question 1 ©NESA 2021 SI Q3 ●●●

Which of the following is **not** a fundamental particle in the Standard Model of matter?

A Electron

B Gluon

C Muon

D Proton

Question 2 ●●●

Particle accelerators can manipulate particles to allow scientists to study the structure of matter. Which of the following particles **cannot** be used for this purpose?

A Electrons

B Photons

C Protons

D Lead nuclei

Question 3 ©NESA 2019 SI Q12 ●●●

The table shows two types of quarks and their respective charges.

Quark	Symbol	Charge
Up	u	$+\dfrac{2}{3}$
Down	d	$-\dfrac{1}{3}$

In a particular nuclear transformation, a particle with a quark composition udd is transformed into a particle with a quark composition uud.

What is another product of this transformation?

A Electron

B Neutron

C Positron

D Proton

Question 4 ◯◯▮

Particle accelerators use fields to manipulate particles. Which of the following correctly describes the use of these fields?

A Superconducting magnets produce fields that increase the momentum of the particles, whereas variations in electric fields accelerate them.

B Superconducting magnets produce fields that change the direction of the particles, whereas variations in electric fields increase their energy.

C Superconducting magnets produce fields that increase the energy of particles, whereas variations in electric fields change their direction.

D Superconducting magnets produce fields that increase the energy of the particles, whereas variations in electric fields increase their momentum.

Question 5 ◯◯▮

Which of the following correctly classifies the subatomic particles listed?

	Lepton	**Hadron**	**Boson**
A	Muon	Neutron	Gluon
B	Gluon	Muon	Neutron
C	Neutron	Gluon	Muon
D	Muon	Proton	Neutrino

Section II: Short-answer questions

> **Instructions to students**
> - Answer all questions in the spaces provided.

Question 6 (2 marks) ◯▮▮

Why is the statement 'protons and neutrons are fundamental particles' incorrect?

Question 7 (5 marks)

a ◯▮▮ Identify a subatomic particle other than a proton, neutron or electron, and name one property of that particle.

2 marks

b ◐◐ For the particle you identified in part **a**, describe a piece of evidence that suggests its existence.

3 marks

Question 8 (7 marks) ©NESA 2018 SII Q34e ◐◐

Using the Standard Model, analyse the roles of both forces and particles in the current understanding of the atom.

Question 9 (6 marks) ██

ANSTO uses a cyclotron to produce ^{18}F. They do this by accelerating protons and bombarding them into ^{18}O, as shown in the diagram below.

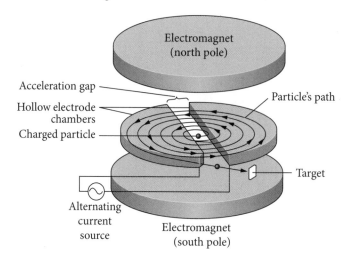

a Write the nuclear equation for this if the products include a neutron and gamma radiation.   2 marks

b Why is it necessary to accelerate the protons to such high speeds?   1 mark

c Outline how the protons are accelerated by the cyclotron.   3 marks

Physics

PRACTICE HSC EXAM 1

General instructions
- Reading time – 5 minutes
- Working time – 3 hours
- Write using black pen
- Draw diagrams using pencil
- Calculators approved by NESA may be used
- A data sheet, formulae sheet and Periodic Table are provided at the back of this paper

---

Total marks: 100

Section I – 20 marks
- Attempt Questions 1–20
- Allow about 35 minutes for this section

Section II – 80 marks
- Attempt Questions 21–35
- Allow about 2 hours and 25 minutes for this section

9780170465298

Section I

20 marks
Attempt Questions 1–20
Allow about 35 minutes for this section

Circle the correct multiple-choice option for Questions 1–20.

Question 1

Blue light is shone on a metal in a photoelectric tube. A small current is detected. Which of the following will cause the current to increase?

A   Decreasing the intensity of the light source

B   Increasing the intensity of the light source

C   Changing the colour to violet

D   Changing the colour to green

Question 2

In the late 1800s, cathode rays were thought to be either a form of electromagnetic radiation or a form of particle. Whose experimental work provided definitive evidence of the nature of cathode rays?

A   Hertz

B   Rutherford

C   Planck

D   J.J. Thomson

Question 3

A current-carrying wire experiences a force when it is placed in a magnetic field. Which of the following shows the correct relationship between force, current and magnetic field?

	Force	Current	Magnetic field
A	→	⊙	↑
B	↑	⊗	→
C	→	↓	⊙
D	⊗	↓	←

Question 4 ©NESA 2015 SIA Q4 ●●

A projectile is launched from a cliff top. The dots show the position of the projectile at equal time intervals.

Assuming negligible air resistance, which diagram best shows the path of the projectile?

A

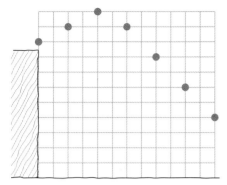

B

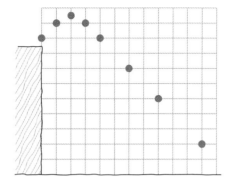

C

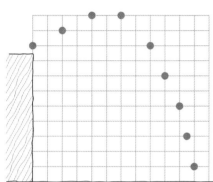

D

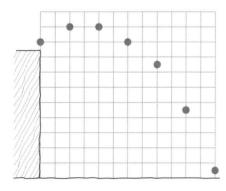

Question 5 ○●

Which of the following was **not** part of Maxwell's work on electromagnetism?

A He was able to show mathematically that light is a consequence of the interaction between electric fields and magnetic fields.

B He provided experimental evidence that electromagnetic waves have a finite speed.

C He provided mathematical models to explain Michael Faraday's work.

D He predicted that electric fields can travels as waves.

Question 6 ●●

A proton is fired in between two electric plates as shown in the diagram. The initial direction of motion is parallel to the plates.

An electron is then fired into the same field, with the same speed, from the same initial position.

The electron, compared with the proton, will have

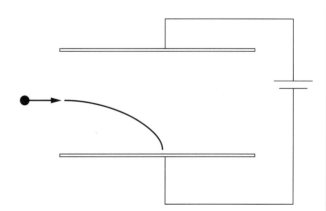

A the same trajectory.

B the same range but will curve upwards.

C a longer range and will also curve downwards.

D a shorter range but will curve upwards.

Question 7 ⦿⦿⦁

A satellite is placed into a geostationary orbit around Earth. NASA decides to place another geostationary satellite around a planet whose mass is twice that of Earth and that has a day length of 48 hours.

How does the orbital radius of the new satellite compare with that of the satellite orbiting Earth?

A It is eight times the size.

B It is double the size.

C It is the same size.

D It is half the size.

Question 8 ⦿⦿⦁

What is the kinetic energy of photoelectrons emitted when ultraviolet light at 200 nm shines on a metal that has a work function of 4.98 eV?

A 6.21 eV

B 1.23 eV

C 4.98 eV

D The photon energy is too low for photoelectrons to be emitted.

Question 9 ⦿⦿⦁

A conical pendulum is set up with radius r, mass M and frequency f Hz. Its angle is measured. A second pendulum is now set up with radius $2r$, mass $2M$ and frequency $\dfrac{f}{2}$ Hz.

What is the ratio of $\tan\theta_r : \tan\theta_{2r}$?

A 1 : 1

B 1 : 2

C 2 : 1

D 4 : 1

Question 10 ©NESA 2021 SI Q16 ⦿⦿⦁

The Sun has an energy output of 3.85×10^{28} W.

By how much does the Sun's mass decrease each minute?

A 4.28×10^{11} kg

B 2.57×10^{13} kg

C 1.28×10^{20} kg

D 7.70×10^{21} kg

Question 11 ●●

A coin-drop machine is shaped like a funnel. Coins roll down a track on their edge before entering the funnel, and then follow a spiral path before dropping through a hole at the bottom. An example is shown on the right.

A coin is dropped into another coin-drop machine that has an initial radius of 50 cm and an inclined side of 35° to the horizontal. It can be treated as a banked curve.

What is the maximum velocity the coin can be launched at so that it remains in the funnel in the first turn?

A $2.3 \, \text{m s}^{-1}$

B $15.2 \, \text{m s}^{-1}$

C $23 \, \text{m s}^{-1}$

D $1.85 \, \text{m s}^{-1}$

EyeEm / Alamy Stock Photo

Question 12   ©NESA 2017 SI Q14   ●●

The diagram shows a DC circuit containing a transformer.

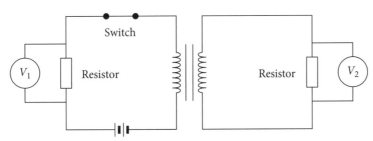

The potential differences V_1 and V_2 are measured continuously for 4 s. The switch is initially closed. At $t = 2$ s, the switch is opened.

Which pair of graphs shows how the potential differences V_1 and V_2 vary with time over the 4-second interval?

A

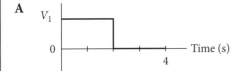
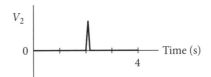

B

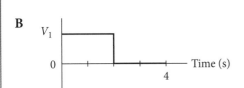
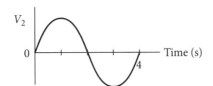

C

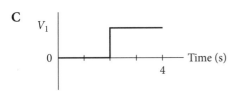
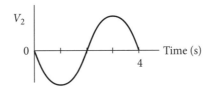

D

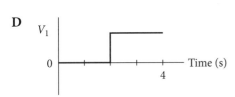
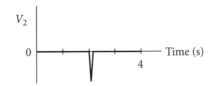

Question 13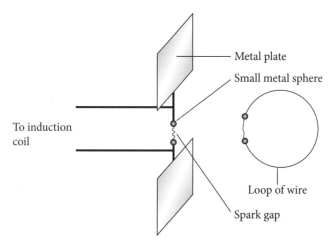

In a hypothetical twin-paradox scenario, Toni and Alex are twins. When they are aged 25, Toni leaves for Alpha Centauri, 4.3 light-years away, and returns, travelling at 0.6c each way.

What will Toni and Alex's ages be upon Toni's return?

	Toni	Alex
A	36	39
B	30	32
C	39	36
D	32	30

Question 14

Hertz was the first to show the existence of radio waves, using the set-up shown in the diagram below.

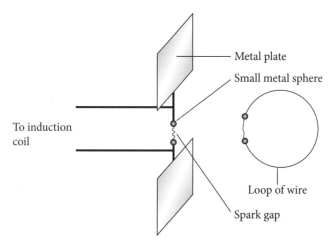

A spark was produced in the loop of wire, which was not connected to the induction coil.

The spark from the induction coil was produced by a wave of $f = 50\,\text{MHz}$. Hertz calculated the wavelength of the invisible radio wave from the standing pattern that was formed. The pattern is shown below.

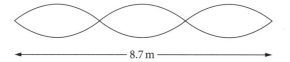

From this information, determine the speed of the radio wave as measured by Hertz.

A $3.0 \times 10^8\,\text{m s}^{-1}$

B $2.9 \times 10^8\,\text{m s}^{-1}$

C $5.8\,\text{m s}^{-1}$

D $6.0\,\text{m s}^{-1}$

Question 15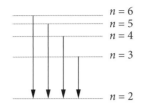

The hydrogen emission spectrum shown below represents the Balmer series.

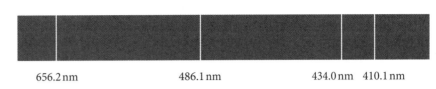

Balmer series (visible light)   **The hydrogen emission spectrum**

Which statement best describes how this supports Bohr's model of the atom involving the discrete energy of electrons?

A   When electrons move between energy levels, they release energy with a specific wavelength. The arrows represent that movement, and the larger the arrow, the larger the wavelength.

B   When electrons receive a specific energy, they move to different energy levels, from higher levels to lower levels.

C   When electrons move between energy levels, they release energy with a specific wavelength. The arrows represent that movement, and the larger the arrow, the smaller the wavelength.

D   When electrons move between energy levels, they release energy with a specific wavelength. The arrows represent that movement. Electrons that moved from $n = 6$ would release more energy than electrons that moved from $n = 4$.

Question 16

At a rubbish processing plant, the metallic materials such as copper and aluminium are separated using magnets.

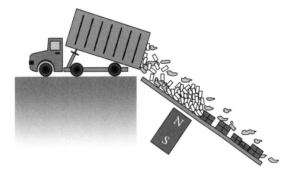

This works because

A   metals will stick to the magnet as they move close to it.

B   eddy currents generated in the metals will cause metals to be attracted to the magnets.

C   a magnetic field is produced in the metals that will be repelled by the magnets and thus make them move faster.

D   eddy currents will cause the metallic materials to move more slowly than other materials.

Question 17 ⬤⬤⬤

Which of the following best compares the characteristics of stars in region A to those in D in the Hertzsprung–Russell diagram below?

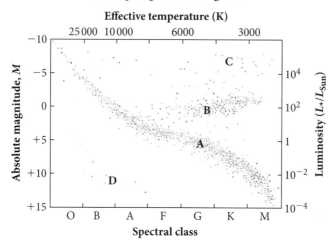

	Star A	Star D
A	Larger radius	Smaller radius
B	Cooler	Hotter
C	Cooler, larger radius	Hotter, smaller radius
D	Cooler, larger radius, shorter life span	Hotter, smaller radius, longer life span

Question 18 ©NESA 2020 SI Q19 ⬤⬤⬤

A conductor *PQ* is in a uniform magnetic field. The conductor rotates around the end *P* at a constant angular velocity.

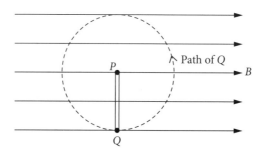

Which graph shows the induced EMF between *P* and *Q* as the conductor completes one revolution from the position shown?

A

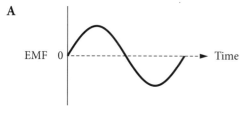

B

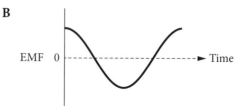

C

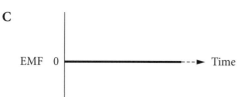

D

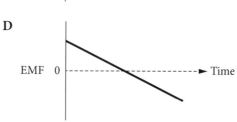

Question 19 ⬤⬤⬤

A beam of electrons travelling at $0.4c$ is fired at two slits. A diffraction pattern is observed on a detector 1 m away, with the first maximum being 1 mm from the centre.

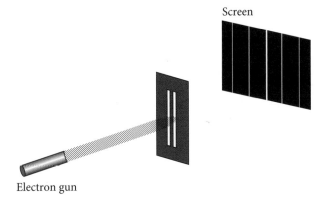

Screen

Electron gun

Not to scale

What is the separation between the slits?

A 1.2×10^{-22} m

B 6.1×10^{-9} m

C 5.6×10^{-9} m

D 5.6×10^{-12} m

Question 20 ©NESA 2019 SI Q19   ⬤⬤⬤

Consider the following nuclear reaction.

$$W + X \rightarrow Y + Z$$

Information about W, X and Y is given in the table.

Species	Mass defect (u)	Total binding energy (MeV)	Binding energy per nucleon (MeV)
W	0.002 388 17	2.224 566	1.112 283
X	0.009 105 58	8.481 798	2.827 266
Y	0.030 376 64	28.295 66	7.073 915

Which of the following is a correct statement about energy in this reaction?

A   The reaction gives out energy because the mass defect of Y is greater than that of either W or X.

B   It cannot be deduced whether the reaction releases energy because the properties of Z are not known.

C   The reaction requires an input of energy because the mass defect of the products is greater than the sum of the mass defects of the reactants.

D   Energy is released by the reaction because the binding energy of the products is greater than the sum of the binding energies of the reactants.

Section II

80 marks
Attempt Questions 21–35
Allow about 2 hours and 25 minutes for this section

Instructions
- Answer the questions in the spaces provided. These spaces provide guidance for the expected length of response.
- Show all relevant working in questions involving calculations.
- Extra writing space is provided at the back of this booklet. If you use this space, clearly indicate which question you are answering.

Question 21 (5 marks)

UV light is shone on the surface of a piece of zinc metal.

a If the UV light has a frequency of 8.0×10^{14} Hz, determine the energy per photon.   2 marks

b A student decided to test different light sources to see when the photoelectric effect can be seen. Below is a diagram showing how they tested this on an electroscope.

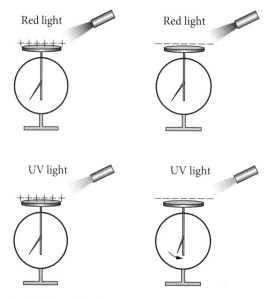

Explain what can be concluded from the observations.   3 marks

End of Question 21

Question 22 (4 marks)

A mass spectrometer separates different ions with masses m_1 and m_2. They are passed through a magnetic field, and the position at which they strike the plate is recorded.

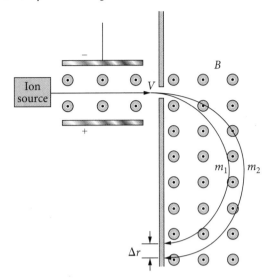

a ◐ How does m_1 compare to m_2?   1 mark

_____

_____

b ◑ Derive an expression for Δr.   3 marks

_____

_____

_____

Question 23 (5 marks) ◑

Using a named example, explain how a particle accelerator relies on the physics of electromagnetism and relativity to probe the structure of matter.

_____

_____

_____

_____

_____

_____

_____

_____

_____

_____

_____

End of Question 23

Question 24 (4 marks)

A common model to demonstrate polarisation is one where a rope is passed through a fence, as shown in the diagram below.

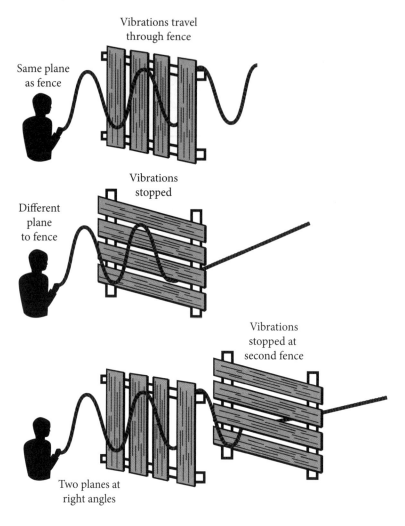

Discuss the strengths and weaknesses of this model.

End of Question 24

Question 25 (7 marks)

An astronomer examines a distant star and can determine the period and radius of several planets around that star.

After collecting the data, the astronomer decided to record the square of the period and cube of the radius.

Period squared $(\times 10^{13}\,s^2)$	Radius cubed $(\times 10^{32}\,m^3)$
1.22	4.13
3.36	11.3
7.14	24.1
9.80	33.1
13.0	44.0

a   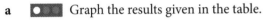 Graph the results given in the table.   2 marks

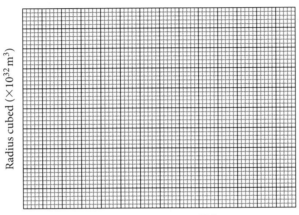

b   Assess the validity of the astronomer's results in determining the relationship between radius and period.   3 marks

c   Using the graph, determine the mass of the star.   2 marks

End of Question 25

Question 26 (5 marks)

In the circus act 'Globe of Death', shown below, a motorcyclist can ride up the globe almost to the 'equator', seeming to defy gravity.

Richard Gosling / Newspix

a   Explain why this is possible.   3 marks

_____

_____

_____

_____

_____

_____

b   The combined mass of the motorcycle and rider is 350 kg, and the diameter of the Globe is 4.86 m. Determine the minimum speed of the motorcyclist to complete a vertical loop of the Globe.   2 marks

_____

_____

_____

_____

End of Question 26

Question 27 (4 marks) ⚫⚫⚫

An electric motor was connected to a power source, and the speed and voltage across the motor were recorded as it began to spin. The data shown in the graph below was collected.

Explain in terms of physics principles what causes the trends in this graph.

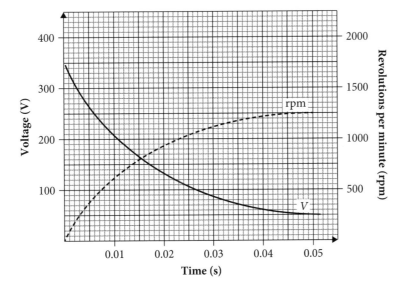

End of Question 27

Question 28 (4 marks)   ©NESA   2020 SII Q27   ●●●

The following apparatus is used to investigate light interference using a double slit.

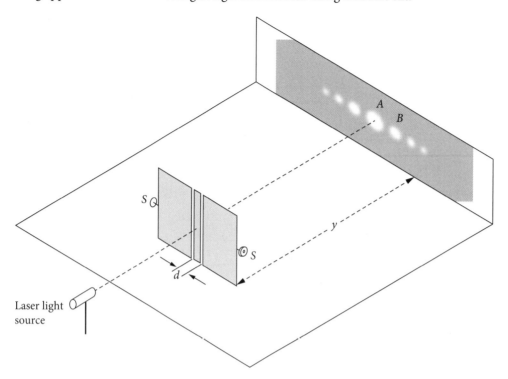

The distance, y, from the slits to the screen can be varied. The adjustment screws (S) vary the distance, d, between the slits. The wavelength of the laser light can be varied across the visible spectrum. The diffraction pattern shown is for a specific wavelength of light.

Explain **two** methods of keeping the distance between the maxima at A and B constant when the wavelength of the laser light is reduced.

End of Question 28

Question 29 (5 marks)

Nuclear fission is known to release energy. One example is the fission of ^{235}U when a neutron is captured by the nucleus.

$$^{235}_{92}U + ^{1}_{0}n \rightarrow ^{141}_{56}Ba + ^{92}_{36}Kr + 3^{1}_{0}n$$

Particle	Mass (u)
^{235}U	235.0439
^{141}Ba	140.9144
^{92}Kr	91.9263
Neutron	1.008 665

The rate of this reaction can be controlled, as in a reactor, or uncontrolled, as in a bomb, as shown below.

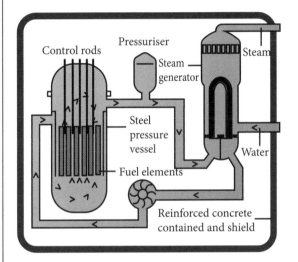

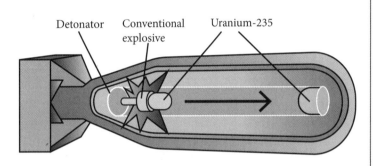

Compare controlled and uncontrolled fission reactions in terms of the process and the energy produced.

End of Question 29

Question 30 (5 marks)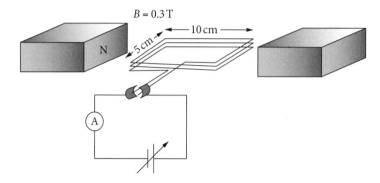

A student constructs a simple motor, as shown in the diagram below.

$B = 0.3\,\text{T}$

N 5 cm 10 cm

A

It starts to turn when the student applies a 1.5 A current.

a   Determine the magnitude of the initial torque on the motor.   2 marks

b   The student wishes to modify the motor so that it works on AC supply. What alterations would be necessary and how might the change to AC alter the motor's characteristics?   3 marks

End of Question 30

Question 31 (7 marks) ●●●

A student performs two experiments, A and B.

In experiment A, shown in the diagram below, the student places a positively charged Styrofoam ball attached to a metal disc that is free to rotate. Using a separate Styrofoam ball, also positively charged, the student causes the wheel to rotate.

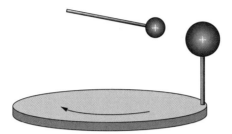

In experiment B, shown below, the student moves a magnet across the disc, which causes the disc to rotate.

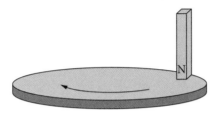

Using physics principles, compare the forces involved in each experiment and how the student makes the disc rotate.

End of Question 31

Question 32 (5 marks) ©NESA 2021 SII Q32 ●●●

Two students perform an investigation with a piece of elastic laid out straight on a table. The elastic is fixed at one end and has three markings at regular intervals. The distances from each marking to the fixed end, d_1, d_2 and d_3, are measured as shown.

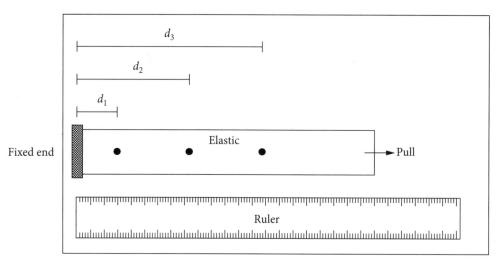

Top view of table

A student pulls the elastic to extend it, and the new values of d_1, d_2 and d_3 are measured. The student observes that each value has doubled.

How well do the observations from this investigation model the evidence that led to Hubble's discovery of the expansion of the Universe? Justify your answer.

End of Question 32

Question 33 (5 marks)

In 1913, Niels Bohr was able to develop a revised model of the atom that could explain the production of emission spectra. However, he could not explain the 'discrete orbits' that were part of his model.

Describe the appropriate mathematical models used by de Broglie to advance the model proposed by Bohr.

End of Question 33

Question 34 (8 marks) ⬤⬤⬤

A student designs a machine to lift heavy objects. They attach a motor to a lever as shown in the diagram below.

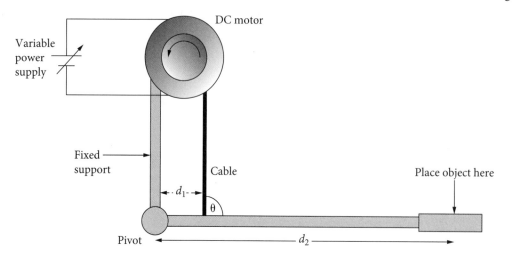

Explain the factors that limit the amount of mass that can be lifted by this device. Use mathematical models to support your answer.

End of Question 34

Question 35 (7 marks) ©NESA 2021 SII Q34

A 3.0 kg mass is launched from the edge of a cliff.

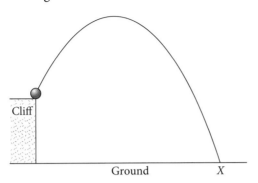

The kinetic energy of the mass is graphed from the moment it is launched until it hits the ground at X.
The kinetic energy of the mass is provided for times t_0, t_1 and t_2.

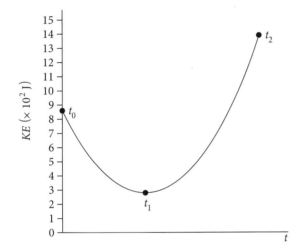

Time, t	KE (J)
t_0	864
t_1	284
t_2	1393

a ⬤⬤ Account for the relative values of kinetic energy at t_0, t_1 and t_2.   4 marks

_____

_____

_____

_____

_____

_____

_____

_____

_____

_____

_____

Question 35 continues on page 146

b   ▢▢▢ The horizontal component of the velocity of the mass during its flight is $13.76\,\mathrm{m\,s^{-1}}$.
Calculate the time of flight of the mass.   3 marks

END OF PAPER

SECTION II EXTRA WORKING SPACE

If you use this space, clearly indicate which question you are answering.

SECTION II EXTRA WORKING SPACE

If you use this space, clearly indicate which question you are answering.

Physics

PRACTICE HSC EXAM 2

General instructions
- Reading time – 5 minutes
- Working time – 3 hours
- Write using black pen
- Draw diagrams using pencil
- Calculators approved by NESA may be used
- A data sheet, formulae sheet and Periodic Table are provided at the back of this paper

---

Total marks: 100

Section I – 20 marks
- Attempt Questions 1–20
- Allow about 35 minutes for this section

Section II – 80 marks
- Attempt Questions 21–35
- Allow about 2 hours and 25 minutes for this section

Section I

20 marks
Attempt Questions 1–20
Allow about 35 minutes for this section

Circle the correct multiple-choice option for Questions 1–20.

Question 1

An object thrown at an angle of 45° is tracked and its horizontal and vertical displacements are recorded with respect to time.

Which of the following correctly shows the path in the horizontal and vertical directions?

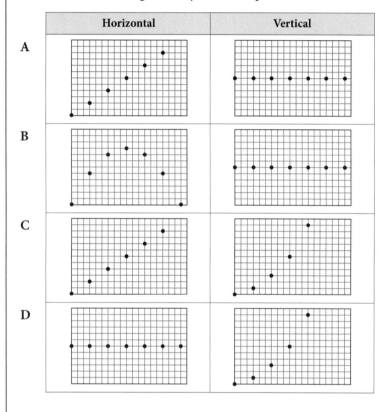

Question 2 ©NESA 2000 SI Q12

A magnet is free to spin. When the switch is closed, the magnet spins in the direction shown.

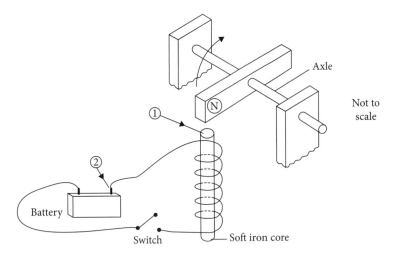

What is the magnetic polarity of the end of the soft iron core marked 1 and the polarity of the battery terminal marked 2?

	Polarity of magnet 1	Polarity of battery 2
A	South	Positive
B	South	Negative
C	North	Positive
D	North	Negative

Question 3

A square loop of wire is placed in a magnetic field as shown in the diagram below. Which of the following changes would cause the greatest amount of EMF to be generated?

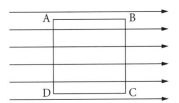

A   Doubling the length of the sides

B   Doubling the strength of magnetic field

C   Rotating the loop by 45° about axis AB

D   Rotating the loop by 45° about axis BC

Question 4

A train travelling at 0.4c turns on its headlight. A person standing in front of the train sees the light turn on. At what speed will they register the light?

A   0.6c

B   1.0c

C   1.4c

D   1.2c

Question 5 ⬤▢▢

Particle accelerators speed up particles. What property does not affect an accelerator's ability to manipulate particles?

A   Mass

B   Volume

C   Speed

D   Charge

Question 6 ⬤▢▢

Newton and Huygens had opposing views about the nature of light. Which statement is correct regarding their views?

A   Newton believed that light could travel as a wave or a particle, depending on the experiment, whereas Huygens believed that light only travelled as waves.

B   Newton believed that light could travel only as a particle, whereas Huygens believed that light could travel as a wave or particle, depending on the experiment.

C   Newton believed that light could travel only as a wave, whereas Huygens believed that light could travel only as a particle.

D   Newton believed that light could travel only as a particle, whereas Huygens believed that light could travel only as a wave.

Question 7 ©NESA 2021 SI Q5 (ADAPTED) ⬤▢▢

The spectrum of an object is shown.

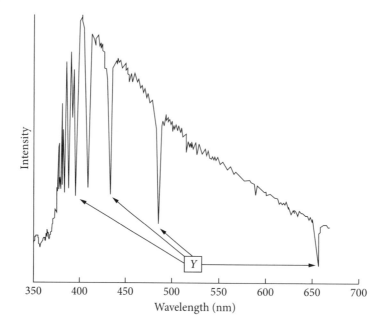

Which row of the table correctly identifies the most likely source of the spectrum and the features labelled *Y*?

	Source of spectrum	Features labelled *Y*
A	Reflected sunlight	Absorption lines
B	Discharge tube	Absorption lines
C	Reflected sunlight	Emission lines
D	Discharge tube	Emission lines

Question 8

Maxwell was able to take the work that was done by many scientists and unify them into four equations that changed physics. Which of the following is **not** a consequence of the equations?

A   The electromagnetic spectrum

B   Electric fields give rise to magnetic fields, which in turn give rise to electric fields

C   Light travels at a constant speed of $3 \times 10^8\,\mathrm{m\,s^{-1}}$ in a vacuum

D   Light is a wave

Question 9

An electron is moving through an area that has an electric and magnetic field. It travels in a straight line.

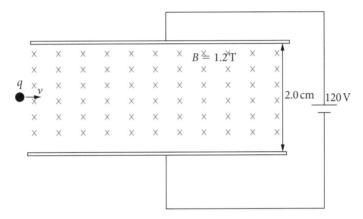

What is the speed of the electron?

A $5.0 \times 10^3\,\mathrm{m\,s^{-1}}$

B $5.0 \times 10^1\,\mathrm{m\,s^{-1}}$

C $2.9 \times 10^2\,\mathrm{m\,s^{-1}}$

D $2.9\,\mathrm{m\,s^{-1}}$

Question 10

The diagram below shows a step-down transformer. The coils in both the primary and secondary coils can be seen.

The ratio of the primary voltage to the secondary voltage is

A $12:18$

B $18:6$

C $2:3$

D $3:2$

Question 11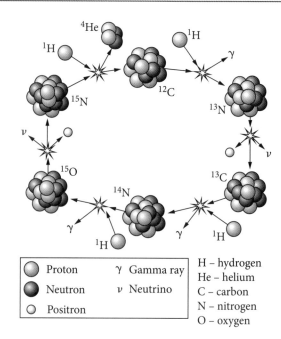

The diagram represents the reactions that occur in the CNO cycle.

Which of the following statements is correct?

A ^{12}C is a product in the net reaction.

B ^{4}He is produced from the cycle.

C   Energy must be added for the reaction to occur.

D   No new elements are produced during the cycle.

○ Proton	γ Gamma ray
● Neutron	ν Neutrino
○ Positron	

H – hydrogen
He – helium
C – carbon
N – nitrogen
O – oxygen

Question 12

The graph below shows the measured black body curve for the Sun.

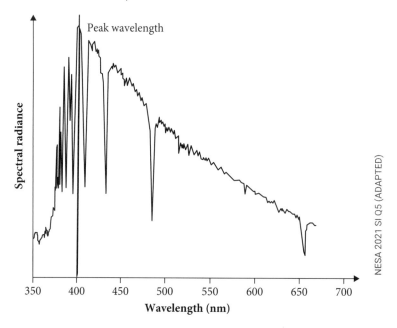

NESA 2021 SI Q5 (ADAPTED)

If we were to look at the spectra of an O-type Main Sequence star compared to this, which statement would be correct about the peak?

	Wavelength of O-type star	Spectral radiance of O-type star
A	Shorter	Lower
B	Longer	Lower
C	Shorter	Higher
D	Longer	Higher

Question 13 ◼◼▨

Below are an image of an AC generator and graphs of the EMF and flux produced.

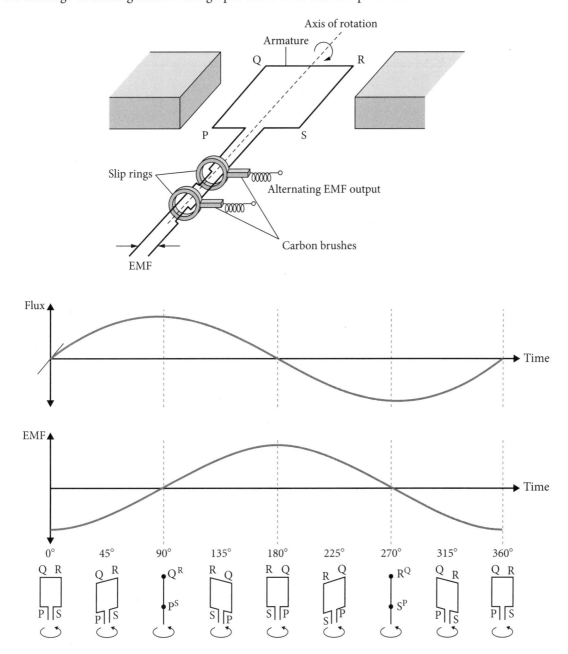

If we were to use a DC generator, how would the graphs change?

A   The EMF peak would be shifted 90°.

B   The flux peaks would all be positive.

C   The EMF would all be positive, but the flux would remain the same.

D   The peaks for EMF and flux would appear at the same point.

Question 14 ©NESA 2015 SIA Q14 ●●

A passenger is playing billiards on a train that is travelling forwards on a level track. The ball takes the path shown when hit by the cue.

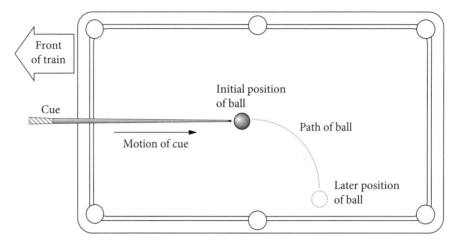

What can be inferred about the motion of the train?

A It is turning left.

B It is speeding up.

C It is turning right.

D It is slowing down.

Question 15 ●●

According to Bohr's atomic model, a photon with a specific energy may, or may not, move an electron from $n = 3$ to $n = 5$.

Which of the following energies would allow for this to occur?

A 1.60×10^{-19} J

B 1.50×10^{-19} J

C 4.50×10^{-19} J

D None of these would cause an electron to move from $n = 3$ to $n = 5$.

Question 16 🔘🔘🔲

Two coils are placed next to each other as shown in the diagram below.

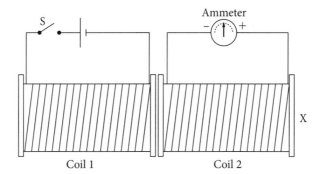

When the switch S is closed, which statement is true about the direction of the magnetic field at X and the initial movement of the needle of the ammeter?

	B direction at X	**Needle direction**
A	Left	Left
B	Left	Right
C	Right	Left
D	Right	Right

Question 17 ©NESA 2020 SI Q18 🔘🔘🔘

An observer sees Io complete one orbit of Jupiter as Earth moves from P_1 to P_2 and records the observed orbital period as t_P. Similarly, the time for one orbit of Io around Jupiter was measured as Earth moved between the pairs of points at Q, R and S, with the corresponding measured periods of Io being t_Q, t_R and t_S.

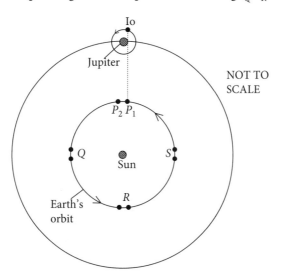

Which measurement of the orbital period would be the longest?

A t_P

B t_Q

C t_R

D t_S

Question 18 ⬤⬤⬤

A student sets a small piece of copper on a pole that is free to pivot at P. The copper is set in a swinging motion.

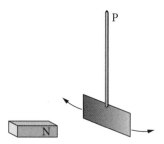

The student then brings a magnet close to the copper and notices that the copper slows down considerably more quickly.

Which of the following diagrams correctly identifies the direction of copper and the eddy currents that are responsible for the braking?

A

B

C

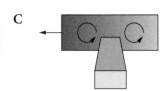

D

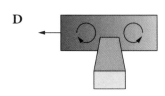

Question 19 ⬤⬤⬤

In a particle accelerator, muons with a rest half-life of 1.56 μs are accelerated to a speed of 0.6c, towards a detector 100 m away. What percentage of the muons will be detected?

A   29%

B   36%

C   78%

D   82%

Question 20 ⬤⬤⬤

The graph below shows the kinetic energy of photoelectrons emitted as a plate made of two metals is exposed to light of varying wavelength.

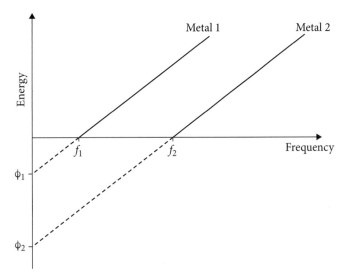

Note that f_2 is three times f_1.

Determine the difference between the energy of the photoelectrons emitted from each metal when the light shone on the plate reaches a frequency of $4f_1$.

A hf_1

B $2hf_1$

C $3hf_1$

D $4hf_1$

Section II

80 marks
Attempt Questions 21–35
Allow about 2 hours and 25 minutes for this section

Instructions
- Answer the questions in the spaces provided. These spaces provide guidance for the expected length of response.
- Show all relevant working in questions involving calculations.
- Extra writing space is provided at the back of this booklet. If you use this space, clearly indicate which question you are answering.

Question 21 (4 marks)

Clare is pushing Riley around on a merry-go-round, with Riley 1.0 m from the centre.

a Determine the force that Clare is exerting if she can produce a torque of 4 N m. 2 marks

b Riley places his toy dinosaur, Rex, at the very edge of the merry-go-round, and says 'Rex is moving faster than me'. Support this statement mathematically. 2 marks

End of Question 21

Question 22 (5 marks)

A projectile was launched, and its trajectory was recorded using video analysis. The data was transposed onto a grid that shows its altitude and range every second. Each grid square side represents 5.0 m.

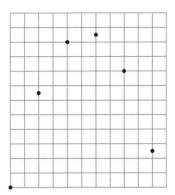

a   ● Determine the horizontal velocity of the projectile.   1 mark

b   ●● Determine the initial velocity in the vertical direction.   2 marks

c   ●● What was the projectile's maximum height?   2 marks

End of Question 22

Question 23 (5 marks) ◯◯▨

Unpolarised light of 100 lux is shone through a polarising filter followed by three other filters at the angles shown.

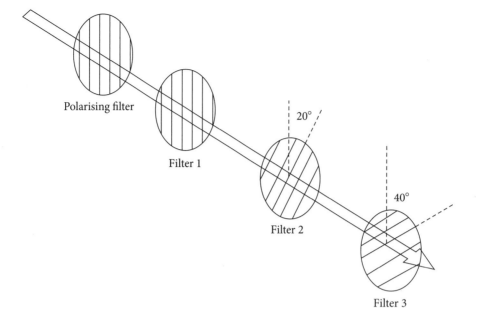

a   Outline how polarisation supports a model of light.   2 marks

b   Determine the intensity of light emerging after the three filters.   3 marks

End of Question 23

Question 24 (5 marks) ©NESA 2016 SIB Q25 ●●

Two teams carried out independent experiments with the purpose of investigating Newton's Law of Universal Gravitation. Each team used the same procedure to accurately measure the gravitational force acting between two spherical masses over a range of distances.

The following graphs show the data collected by each team.

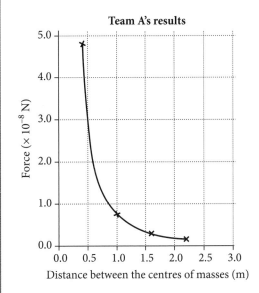

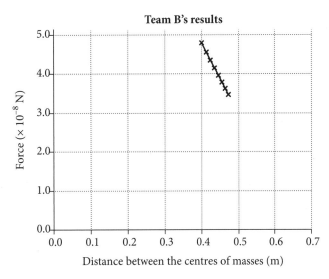

a Compare qualitatively the relationship between force and distance in the graphs.   2 marks

b Assess the appropriateness of Team A's data and Team B's data in achieving the purpose of the experiments.   3 marks

End of Question 24

Question 25 (5 marks) ⬤◻◻

Two current-carrying wires are positioned next to each other as shown. The top wire (wire 1) is held in place and cannot move.

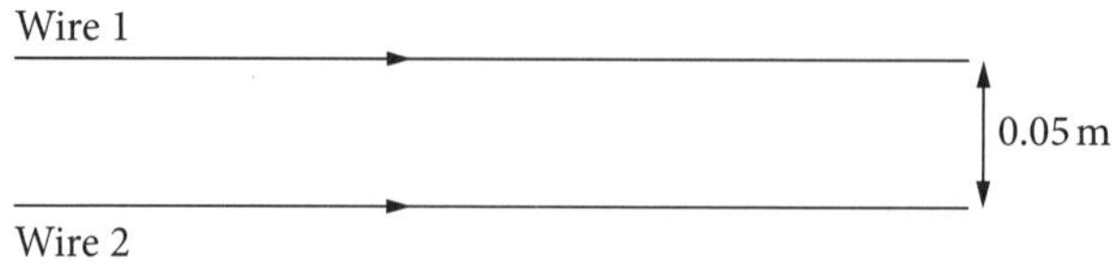

Wire 1 →

0.05 m

Wire 2 →

a The force of attraction between the wires is $8.0 \times 10^{-5}\,\text{N}\,\text{m}^{-1}$. The current in wire 1 is three times the current in wire 2. Determine the current in wire 2.   2 marks

b Where would you position a third wire (wire 3), carrying current I_3, which is the same magnitude as for wire 2?   3 marks

Question 26 (5 marks) ⬤⬤◻

Early models of the atom were soon tested by experiments. The evidence from these experiments demonstrated inadequacies in the models. With reference to some of the models of the structure of the atom from Thomson to Schrödinger, show how these progressed the atomic model.

End of Question 26

Question 27 (8 marks) ©NESA 2019 SII Q31 ●●

A student suspends an electric ceiling fan from a spring balance.

The fan is switched on, reaching a maximum rotational velocity after 10 s.

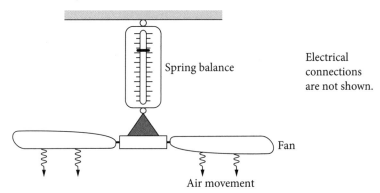

Spring balance

Electrical connections are not shown.

Fan

Air movement

a   Explain the changes that would be observed on the spring balance in the first 15 seconds after the fan is switched on.   4 marks

_____

_____

_____

_____

_____

_____

_____

_____

Question 27 continues on page 166

b   The student predicted that the current through the fan's motor would vary as shown on the graph.

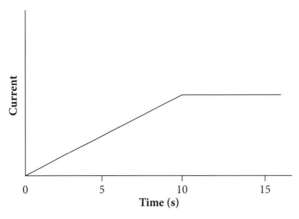

Assess the accuracy of the student's prediction.   4 marks

End of Question 27

Question 28 (4 marks)

Galileo attempted to measure the speed of light by the following experiment. He stood 1.6 km away from another person, and they both held a shuttered lantern. One person flashed the light and the other flashed back as soon as they saw the light. The distance divided by this time gave the speed.

A microwave oven generates standing waves within the metal interior of the oven, where they are absorbed by food. A high-school student attempted to measure the speed of light by heating a piece of chocolate in a microwave oven ($f = 2450$ MHz) after removing the rotating plate. They noticed a series of melted spots in the chocolate, and measured the distance between them.

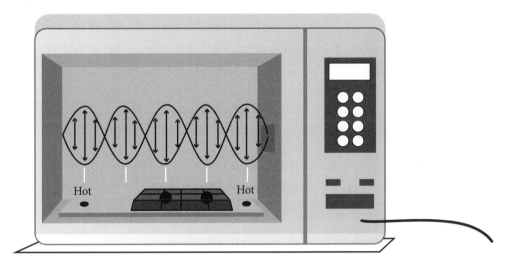

Compare these two methods of measuring the speed of light.

End of Question 28

Question 29 (3 marks) ⬤⬤▪

'Nuclear energy goes against the conservation of energy – you can get more energy out than you put in. You just have to put in a tiny amount of energy to split a nucleus and you get out so much more.'

Assess this statement.

_____

_____

_____

_____

_____

Question 30 (5 marks)

A teacher used five-cent coins to simulate the decay of a radioactive isotope. The students shook the coins onto a piece of paper, and if the tails was face up, it was said to have decayed and was removed.

Each shake represents 1700 years in this simulation. The following data was obtained.

Shake number	Number of coins remaining (not decayed)	Number of coins with tails side up (decayed)
0	304	0
1	152	152
2	62	90
3	38	24
4	17	21
5	14	3
6	11	3
7	8	3
8	5	3
9	3	2
10	1	2

a   ⬤⬤▪ Calculate the hypothetical time that has passed if only 22 coins remain.   3 marks

_____

_____

_____

_____

_____

_____

Question 30 continues on page 169

b ◖○◗ Discuss a limitation of this model. 2 marks

Question 31 (6 marks)

Galileo was able to observe the moons of Jupiter as they revolved around the planet.

7 January 1610		* ○ *
8 January		○ * * *
9 January		
10 January	* * ○	
11 January	* * ○	
12 January		* *○ *
13 January		* ○***

Galileo's observations of the moons (labelled as *) around Jupiter (labelled as a circle) over a number of days in 1610.

Over a period of days, Galileo noted the position of Ganymede and recorded its position in terms of Jupiter diameters or JD. In other words, 2 JD is equivalent to two Jupiter diameters = 2 × 139 820 km.

Galileo was able to determine the mass of Jupiter by plotting the positions of Ganymede over several days, as shown in the graph below.

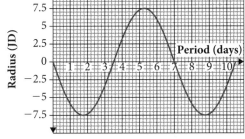

a ◖○◗ Determine the period and the radius of Ganymede's orbit. 2 marks

b ◖○◗ Calculate the mass of Jupiter. 2 marks

Question 31 continues on page 170

c   ▮▮▮ If you were to measure the positions of another moon of Jupiter, discuss how the appearance of the graph might change.

2 marks

Question 32 (4 marks) ©NESA 2019 SII Q35 ●●●

The apparatus shown is attached horizontally to the roof inside a stationary car. The plane of the protractor is perpendicular to the sides of the car.

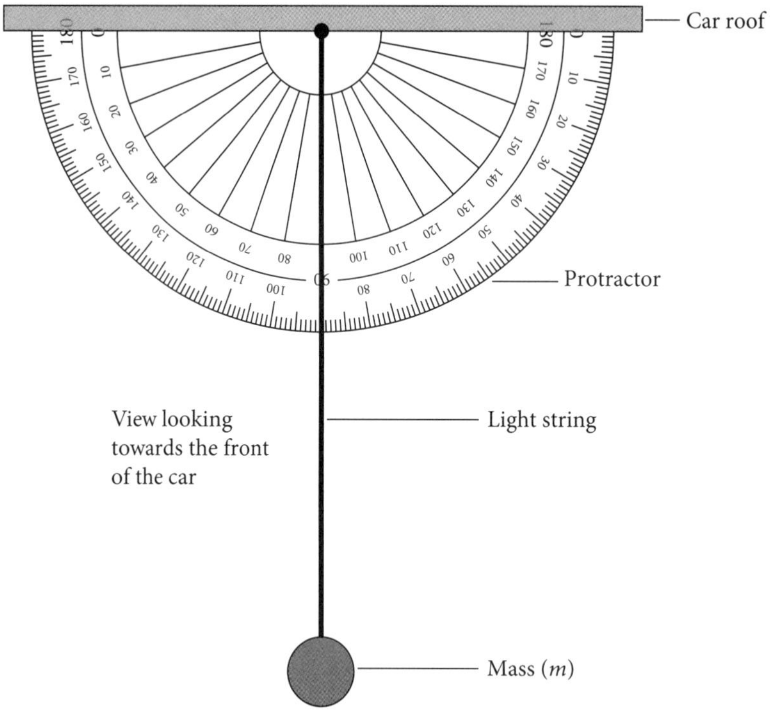

The car was then driven at a constant speed (v), on a horizontal surface, causing the string to swing to the right and remain at a constant angle (θ) measured with respect to the vertical.

Describe how the apparatus can be used to determine features of the car's motion. In your answer, derive an expression that relates a feature of the car's motion to the angle θ.

End of Question 32

Question 33 (9 marks) 🔘🔘▪️

'The black body curve not only helps our understanding of the world around us, and beyond, but was instrumental in completely altering our understanding of physics.'

Discuss this statement.

End of Question 33

Question 34 (6 marks)

A particle with mass m and charge $+q$ enters midway between two plates that are separated by distance d.
A voltage V is applied across the plate.

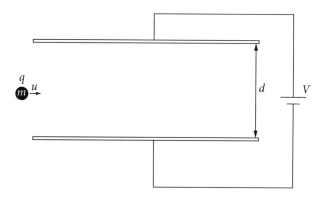

a   ::: Show that the time it takes for the charge to reach a plate is $t = d\sqrt{\dfrac{m}{qV}}$.   3 marks

b   ::: The charge is now projected at an angle of 45° from the base of the plate.

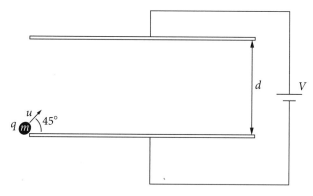

Derive an expression for the range of the particle.   3 marks

End of Question 34

Question 35 (6 marks) ⬤⬤⬤

In 1964, Peter Higgs hypothesised the existence of a Higgs field, which could explain why particles, such as quarks, have mass. The field would have an associated boson, called a Higgs boson, and it was predicted to have a mass of around $125\,\text{GeV}/c^2$.

CERN constructed the Large Hadron Collider to detect this boson. It started operation in 2008.

In 2012, the teams from two detectors on the LHC ring, ATLAS and CMS, announced that they had detected, independently, the boson, with mass of $125\,\text{GeV}/c^2$.

Using the statement above as an example, examine the role of particle accelerators in obtaining evidence that tests and/or validates aspects of the Standard Model of matter.

END OF PAPER

SECTION II EXTRA WORKING SPACE

If you use this space, clearly indicate which question you are answering.

SOLUTIONS

Test 1: Projectile motion

Multiple-choice solutions

Question 1

B equal to g

> The moment the projectile leaves the aircraft, the only force acting on the projectile is mg. This is constant, and therefore its acceleration, which is equal to g, will not change as it falls.

Question 2

B decrease the range

> The horizontal velocity of an object only affects its horizontal motion, because the horizontal and vertical components of velocity are independent of each other. This means if the horizontal velocity is decreased, it will not travel as far (i.e. its range will be less).
>
> Although the horizontal velocity has changed, the height is constant and so the time of flight would be the same because the object experiences the same acceleration, g. (Hint: Imagine observing from a position such that the marble is rolling directly towards you. You would see only the vertical motion.)

Question 3

C The magnitude of the acceleration at position 2 is the same as at position 5.

> After launch, the only force on the ball is the force due to gravity. Therefore, the magnitude of its acceleration is constant. **A** is not correct because the direction of motion is different at position 5 than at position 1. **B** is not correct because at point 3 the net force on the ball is still equal to the force due to gravity. **D** is not correct – there is no horizontal acceleration because there is no horizontal force and the parabolic path is symmetrical; therefore, the horizontal velocity is constant.

Question 4

B Jeri's ball will reach the target first because its vertical velocity will be less.

> A smaller vertical component of velocity means a greater horizontal component of velocity. This decreases the time in flight.
>
> **A** is not correct because the acceleration is constant at g. **C** is not correct because the higher throw means the ball will be in the air longer. **D** is not correct because, although the balls have the same speed, the angle will change its horizontal and vertical components. If the horizontal component is increased because the ball is thrown at a lower angle (Jeri), it will take less time for the ball to reach the target.

Question 5

B At the bullseye

The dart and the target will fall at the same rate because they both have the same acceleration, g. Therefore, to hit the bullseye, the dart gun would need to be aimed at the bullseye.

Question 6

A Horizontal velocity: D; vertical velocity: A

The horizontal velocity is constant because there is no horizontal acceleration, so the graph would be a horizontal line. The horizontal velocity is in the positive direction, so the line would be above zero. Therefore, Graph D is the best representation.

The vertical velocity changes at a constant rate of $-9.8\,\text{m s}^{-2}$. This is a straight line on a velocity–time graph. It is a negative value, so the graph slope would be negative. Graph A is the best representation.

Question 7

A Height: decreases; range: decreases

The angle and initial velocity of both hits are the same, so the vertical and horizontal components of the initial velocity would be the same, too.

However, the acceleration from gravity of Earth, g_{Earth}, is six times that of the Moon, g_{Moon}. Hence, the time of flight would be less on Earth, decreasing the maximum height the golf ball would reach and its range.

Question 8

A $6:1$

The formula that combines the known values is $v = u + at$. In determining the time taken to reach the top of the flight, v and u are the same. Because the acceleration is approximately 6 times less on the Moon, the time of flight will be 6 times more, so $t_{\text{Moon}} : t_{\text{Earth}}$ is $6:1$.

B is incorrect. Students may choose **B** is they mistakenly think that time has a squared relationship with velocity instead of a squared relationship with displacement. **C** and **D** have the ratio in the incorrect order, and **D** also shows an incorrect value.

Question 9

C An increasing acceleration opposite to the direction of motion

An increasing acceleration opposite to the projectile's direction of motion would cause its range and maximum height to decrease. **A** is incorrect because an *increased* force causes the path with air resistance, since now there is not only the gravitational effect but also a horizontal force to the left. **B** is incorrect because an increasing horizontal acceleration to the left would reduce the range but would result in the same maximum height. **D** is incorrect — a constant force to the left would cause the path to be a straight line because there is a constant force vertically and horizontally.

Question 10

C

Because the object is launched at 45°, the vertical and horizontal components of initial velocity will be equal in magnitude, i.e. $u_x = u_y$

If we make both u_x and $u_y = u$

Vertically:

$$v = -u\,\mathrm{m\,s^{-1}} \qquad u = u\,\mathrm{m\,s^{-1}} \qquad a = g\,\mathrm{m\,s^{-2}} \qquad t = t \qquad s_y = 0$$

$s = ut + (0.5)at^2$

$0 = ut + (0.5)gt^2$ (1)

Horizontally:

$$v_x = u\,\mathrm{m\,s^{-1}} \qquad t = t \qquad s_x = ut,$$

where $R = ut$ (2)

We can substitute equation (2) into equation (1) and solve for R.

$$0 = \frac{R}{t}t + \frac{gt^2}{2}; \text{ therefore, } R = \frac{gt^2}{2} \text{ and thus } R \text{ is proportional to } t^2.$$

A is incorrect because it shows a linear relationship. **B** is incorrect because it shows the relationship of R^2 versus t. **D** is incorrect because it shows an inverse relationship.

Short-answer solutions

Question 11

If the truck is moving at $50\,\mathrm{km\,h^{-1}}$ to the right, initially the ball is also moving at $50\,\mathrm{km\,h^{-1}}$ to the right.

As the ball is thrown up in the air, it has a downwards acceleration due to gravity of $9.8\,\mathrm{m\,s^{-2}}$, but it has no acceleration in the horizontal direction. This means the ball will follow a parabolic path.

- 3 marks: explains the initial horizontal velocity of the ball and that the vertical acceleration will result in a parabolic path
- 2 marks: describes the motion in components
- 1 mark: identifies parts of the motion

Question 12

a

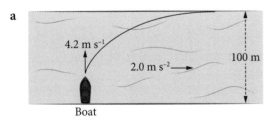

- 1 mark: correctly sketches a parabolic path

b Across the river: Downstream:

$v_x = 4.2 \, \text{m s}^{-1}$ $s_x = 100 \, \text{m}$ $t = ?$ $u = 0 \, \text{m s}^{-1}$

$s_x = vt$ $100 = 4.2(t)$ $t = \dfrac{100}{4.2} = 23.8 \, \text{s}$ $a = 2.0 \, \text{m s}^{-2}$ $s_y = ?$ $t = 23.8 \, \text{s}$

$$s = ut + (0.5)at^2$$
$$= (0.5)(2.0)(t^2)$$
$$= 566 \, \text{m}$$

- 3 marks: correctly calculates the position on the other bank
- 2 marks: makes a correct substitution into a relevant formula
- 1 mark: attempts to calculate the position

Question 13

a $45°$

- 1 mark: identifies the optimum angle

b Vertically: Horizontally:

$v_y = 0$ $v_x = 6.0 \cos 45 \, \text{m s}^{-1}$

$u = 6.0 \sin 45 \, \text{m s}^{-1}$ $s_x = s_x \, \text{m}$

$a = -9.8 \, \text{m s}^{-2}$ $t = 0.866 \, \text{s}$

$t = ?$ $s_x = v_x t$

$v = u + at$ $s_x = 6.0 \cos 45 (0.866)$

$0 = 6.0 \sin 45 - 9.8t$ $= 3.7 \, \text{m}$

$t = 0.433 \, \text{s}$ (time to top of flight)

Therefore time of flight $= 2 \times 0.433$

$$t = 0.866 \, \text{s}$$

This is less than $4.0 \, \text{m}$, so the gazelle would not make it to the other side of the river.

- 3 marks: correctly calculates the distance and determines if the gazelle lands safely based on this calculation
- 2 marks: attempts to calculate and determine if the gazelle lands safely
- 1 mark: makes a substitution into a relevant formula

Question 14

a Any two of the following: no horizontal acceleration (no air resistance); vertical acceleration is constant; horizontal and vertical motion are independent.

- 2 marks: correctly identifies two assumptions that impact the parabolic path
- 1 mark: correctly identifies one assumption that impacts the parabolic path

b Because the gravitational acceleration on the Moon is less than on Earth, the projectile's time of flight would be longer on the Moon. This would result in a larger vertical displacement and a longer range.

- 3 marks: explains two differences between the trajectories
- 2 marks: explains one difference between the trajectories **or** identifies two differences between the trajectories
- 1 mark: identifies one difference between the trajectories

Question 15

a

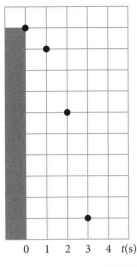

As the horizontal motion is constant, the ball will move equal amounts each time period.

Vertically:

$$s = ut + \frac{1}{2}at^2$$

$$= 0 + \frac{1}{2}(10)(1)$$

$$= 5\,\text{m}$$

This can be done for each time interval to work out the position of the ball at $t = 2$ and $t = 3\,\text{s}$.

- 3 marks: correctly marks the next three positions accurately
- 2 marks: demonstrates constant horizontal velocity **or** marks two correct positions
- 1 mark: marks one position accurately

b $s = ut + \frac{1}{2}at^2$

$$-50 = 0 - \frac{1}{2}(10)t^2$$

$$t = 3.19\,\text{s}$$

- 2 marks: correctly calculates the time to hit the ground
- 1 mark: makes a substitution into a relevant formula

Question 16   ©NESA   2020 MARKING GUIDELINES SII Q24

From the graph, max height is $44\,\text{m}$ and $v_y = 0$ at max height. Therefore:

$$v_y^2 = u_y^2 + 2a_y s_y$$

$$0 = u_y^2 + 2 \times (-9.8) \times 44$$

$$u_y = 29.4\,\text{m s}^{-1}$$

From the graph, time of flight is $6\,\text{s}$. Therefore:

$$u_x = \frac{s_x}{t} = \frac{130}{6} = 21.7\ \text{m s}^{-1}$$

Applying Pythagoras' relationship:

$$u^2 = u_x^2 + u_y^2$$

$$u^2 = 21.7^2 + 29.4^2$$

$$u = 36.5\,\text{m s}^{-1}$$

Calculating angle of launch:

$$\tan\theta = \frac{u_y}{u_x}$$

$$\theta = 54°$$

Therefore $u = 36.5\,\text{m s}^{-1}$, $54°$ above the horizontal

$$= 37\,\text{m s}^{-1}$$

Answers could include:

Calculating u_y using $s_y = u_y t + 0.5a_y t^2$, where $t = 3$ and $s_y = 44\,\text{m}$.

- 4 marks: correctly calculates the magnitude and direction of the initial velocity
- 3 marks: provides the main steps for calculating the initial velocity
- 2 marks: provides some steps for calculating the initial velocity
- 1 mark: provides some relevant information

Test 2: Circular motion

Multiple-choice solutions

Question 1

A 1 : 1

The rate of change of angle is independent of the radius. **B**, **C** and **D** all assume that the radius plays a factor, but that would only be the case if the question were referring to linear velocity.

Question 2

C The acceleration is directed to the east.

Because the car is travelling in a circular path, there must be an acceleration towards the centre of the circle, which in this case means to the east. **A** is incorrect because south is the direction of the car's velocity at the instant shown. **B** is incorrect because west is the direction of the reaction to the centripetal force, often erroneously called the centrifugal force. **D** is not correct because the car is changing direction, and thus velocity, which means it must be accelerating.

Question 3

B

There are two real forces acting, shown in **D**, but the question asks for the net force, which is the resultant of all the forces acting. Therefore, the net force must be only one arrow. The resultant of all the forces acting is the addition of the two vectors in **D**. This net force is the centripetal force, which is shown in **B**.

Question 4

B Line *B*

The only force acting on the mass (in the horizontal direction) is the tension in the string, and the path of the mass is moving up the page. When the string breaks, the mass will continue in an upward path, according to Newton's first law. Path **A** would only occur if the string remained intact. Both **C** and **D** suggest an additional, outwards force, sometimes called the centrifugal force, but this is not a valid force in the frame of reference of the table and is therefore incorrect.

Question 5

A Up the wall

Apart from the centripetal force acting inward, there is also the force from gravity, which is acting downward. Because the people are 'stuck' and are not moving downward, there must be an opposing force to counteract the gravitational force. This is friction, which is acting upward.

B is incorrect because it is referring to the gravitational force. **C** is incorrect because it is referring to the centripetal force. **D** refers to the centrifugal force, which is not appropriate in this frame of reference.

SOLUTIONS – TEST 2

Question 6

A $28\,\mathrm{m\,s^{-1}}$

The linear speed can be determined by $v = 2\pi rf = (2)(3.14)(1.5)(3) = 28\,\mathrm{m\,s^{-1}}$. **B** is incorrect because it has an incorrect substitution of $3.0\,\mathrm{rev\,s^{-1}}$ for the period. **C** is incorrect because it misreads $3.0\,\mathrm{rev\,s^{-1}}$ as the speed, not the frequency. **D** is incorrect because the student has divided the frequency by the radius.

Question 7

B Mg

The centripetal force is provided by the force from gravity on the hanging mass, M. **A** is incorrect because an incorrect substitution has been made for centripetal force, taking $3\,\mathrm{rev\,s^{-1}}$ as velocity. **C** is incorrect because it is the velocity formula squared, not the centripetal force. **D** is incorrect, with the formula having no meaning.

Question 8

B

The car experiences three separate centripetal accelerations: as it enters the loop, as it stays in the loop, and as it leaves the loop. The direction of acceleration as it enters and leaves the loop is outwards, whereas in the loop the acceleration is inward. The radius of the curve the car travels when it is entering and leaving the loop is smaller than the radius of the loop itself, so the acceleration will be greater when it is entering and leaving the loop. **A** is incorrect because it suggests there is no acceleration in the loop. **C** is incorrect because it suggests all accelerations are in the same direction. **D** is incorrect because it does not show the acceleration when the car is entering and leaving the loop.

Question 9

A The magnitude of the net force remains constant.

There are two types of motion occurring, which are independent of each other. The first is projectile motion, where the force applied is the gravitational force, which acts downward and has a constant magnitude. The vertical component of the velocity changes as a result; however, the magnitude of the horizontal velocity remains constant.

The second motion is horizontal uniform circular motion, as a result of the normal force acting on the ball by the surface of the wall, which provides a centripetal force. Because the horizontal speed is constant, the magnitude of the centripetal force is also constant. The vertical motion does not affect the centripetal force.

Therefore, the ball experiences two forces – the gravitational force downward and the centripetal force towards the centre. The sum of these forces is the net force. Because both forces have unchanging magnitudes, the net force must therefore also have a constant magnitude.

B and **C** are incorrect because they only account for one of the forces. **B** ignores the normal force acting on the ball, which is causing the circular motion. **C** ignores the gravitational force acting vertically on the ball. **D** is incorrect because it suggests that the magnitude of the net force changes. Because the magnitude of the normal force and the gravitational force both remain constant, the sum of those forces – the net force – must also remain constant.

Question 10

C 327 N

> To lift the mass, the torque applied by the mass must be equal to the torque applied by hand to the rope. Because the pulley the force is applied to has three times the radius of the pulley the mass is connected to, the force applied will be three times less, thus **C** is correct $\left(\dfrac{mg}{3} = \dfrac{(100)(9.8)}{3} = 327\,\text{N} \right)$.
>
> **A** is incorrect because it suggests the same force has to be applied; in reality, the same torque has to be applied. **B** is incorrect because it incorrectly suggests that the radius doubles. **D** is incorrect because it suggests that the angle for $\tau = Fr\sin\theta$ is 45°; however, the force applied is perpendicular to the radius of the pulley, so the angle is 90°.

Short-answer solutions

Question 11

a $v = \dfrac{2\pi r}{T} = 2\pi rf = 2\pi(0.15)\left(\dfrac{45}{60}\right) = 0.7\,\text{m s}^{-1}$

> - 2 marks: determines the speed
> - 1 mark: makes a partially correct substitution into a correct equation

b For the car to stay on the turntable, $F_c = F_f$

$$\frac{mv^2}{r} = \mu N = \mu mg$$

$$\frac{v^2}{r} = \mu N = \mu g$$

$$\frac{(0.71)^2}{0.15} = \mu(9.8)$$

Therefore, $\mu = 0.34$

> - 3 marks: determines the coefficient of friction
> - 2 marks: makes a partially correct substitution into the correct equations
> - 1 mark: makes a partially correct substitution into a correct equation

c Because $\dfrac{v^2}{r} = \mu g$, and this is independent of mass, a decrease in mass will have no effect on the speed at which it can remain on the turntable.

> - 2 marks: explains why the car with reduced mass will stay on the disc
> - 1 mark: provides relevant information

SOLUTIONS – TEST 2

Question 12

a   The minimum speed will be such that the centripetal acceleration at the top of the loop is equal to the gravitational acceleration.

$$\frac{mv^2}{r} = mg$$

$$\frac{v^2}{r} = g$$

$$\frac{v^2}{0.15} = 9.8$$

Therefore, $v = 1.2\,\text{m s}^{-1}$

- 2 marks: determines v
- 1 mark: makes a partially correct substitution into the correct equation

b   Assuming minimal loss in energy because of friction, the car exits the first loop with the same speed, and therefore the same kinetic energy, as when it entered. Part **a** showed that the required minimum speed for a loop is $v = \sqrt{rg}$ (due to conservation of energy). Thus, the successive loops with a smaller radius each require lower speeds for a successful loop. Because the car makes the first loop, it will easily be able to complete each successive loop because the required minimum speed is less than the previous loop.

- 3 marks: uses the law of conservation of energy and centripetal motion to justify why it would make each successive loop
- 2 marks: attempts to use the law of conservation of energy to justify why it would make each successive loop
- 1 mark: provides relevant information

Question 13

a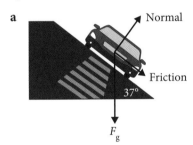

- 2 marks: labels three forces correctly
- 1 mark: labels two forces correctly

b   By equating the forces in the vertical and horizontal components, the following equations are derived.

$$\frac{mv^2}{r} = N\sin\theta$$

$$mg = N\cos\theta$$

$$\frac{\frac{mv^2}{r}}{mg} = \frac{N\sin\theta}{N\cos\theta}$$

$$v^2 = gr\tan\theta$$

$$v = 43\,\text{m s}^{-1}$$

- 2 marks: determines v
- 1 mark: makes a partially correct substitution into the correct equation

Question 14

a

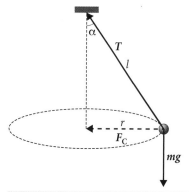

b

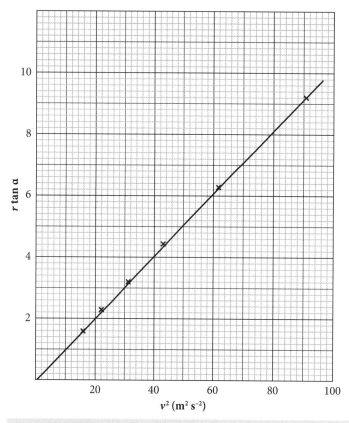

c Because $\tan(\alpha) = \dfrac{\left(\dfrac{mv^2}{r}\right)}{mg}$; therefore, $\dfrac{v^2}{r\tan(\alpha)} = g$

Slope $= \dfrac{1}{g} = \dfrac{(9.0 - 1.0)}{(88 - 10)} = 0.103$; therefore, $g = 9.75\,\mathrm{m\,s^{-2}}$

d The value is close to the accepted value of $9.8\,\mathrm{m\,s^{-2}}$.

Test 3: Motion in gravitational fields

Multiple-choice solutions

Question 1

A

The correct answer is **A** because the force from gravity is the only force acting and so provides the centripetal force. The velocity is tangential to the orbital path. **B** has the force in the wrong direction for this frame of reference (although it would be in this direction in the reference frame of the satellite). **C** has the velocity acting inward, mistaking it for the centripetal acceleration. **D** has the force at a tangent, which is not correct and would not result in circular motion or constant speed.

Question 2

C $7.6\,\text{km s}^{-1}$

$$\frac{GMm}{r^2} = \frac{mv^2}{r}$$

$$\frac{GM}{r} = v^2$$

$$v = \sqrt{\frac{GM}{r}} = \sqrt{\frac{(6.67 \times 10^{-11})(6.0 \times 10^{24})}{(6371 + 547) \times 10^3}} = 7.6 \times 10^3\,\text{m s}^{-1}$$

A is incorrect because it has used g instead of G. **B** is incorrect because it has used the square of r. **D** is incorrect because this is the value of g.

Question 3

A $6371\,\text{km}$

$$mg = \frac{GMm}{r^2}$$

$$g = \frac{GM}{r^2}$$

$$r_E^2 = \frac{GM}{g_E}$$

$$\text{new } r^2 = \frac{G2M}{2g_E} = \frac{GM}{g_E}$$

Therefore, the radius of the planet is the same as Earth, so **A** is correct.

Question 4

D $9:8$

$$F_A : F_B$$

$$\frac{GmM_A}{r_A^2} : \frac{GmM_B}{r_B^2}$$

$$\frac{M}{(2R)^2} : \frac{2M}{(3R)^2}$$

$$\frac{1}{4} : \frac{2}{9}$$

$$9 : 8$$

Therefore, **D** is correct.

Question 5

B The kinetic energy at Y is less than at X.

> The kinetic energy of the satellite is greater the closer it is to the planet. Therefore, the satellite has the greatest kinetic energy at X. **A** is incorrect because total energy does not change. **C** is incorrect because the potential energy increases as the satellite moves away from the planet and potential energy becomes less negative. **D** is incorrect because the area that each sweeps out is the same in the same time according to Kepler's law.

Question 6

A Potential energy: different; orbital speed: same

> The potential energy of any object in orbit is determined by the mass of the planet and the mass of the object as well as the radius of the orbit, according to $U = -\dfrac{GMm}{r}$. Therefore, the piece of dust will have a different amount of gravitational potential energy to the satellite. The orbital speed is only determined by the mass of the planet and its radius as $v = \sqrt{\dfrac{GM}{r}}$; therefore, the piece of dust and the satellite will have the same speed.

Question 7

C

> Potential energy, U, is always negative. When the satellite is in a low Earth orbit, U is constant. However, as it moves to a geostationary orbit, its gravitational potential will increase until it enters a new stable orbit. In this context, increasing potential energy means the potential energy becomes less negative. It is important to note these are graphs of potential energy versus time, not potential energy versus radius.
>
> **A** and **B** are both incorrect because both have U as being positive. **D** is incorrect. Although it is negative, it does not reflect that initially U is constant.

Question 8

D The magnitude of the net force decreases and then increases as the satellite moves from X to Y.

> To work out the change in net force on the satellite, determine the magnitude of the net force at Y and at X.
>
> $$F_Y = F_3M - F_2M$$
> $$= \frac{G3Mm}{(2R)^2} - \frac{G2Mm}{(6R)^2}$$
> $$= GMm\left(\frac{3}{4R^2} - \frac{2}{36R^2}\right)$$
> $$= \frac{GMm}{R^2}\left(\frac{25}{36}\right)$$
>
> $$F_X = F_3M - F_2M$$
> $$= \frac{G3Mm}{(7R)^2} - \frac{G2Mm}{(R)^2}$$
> $$= GMm\left(\frac{3}{49R^2} - \frac{2}{R^2}\right)$$
> $$= \frac{GMm}{R^2}\left(\frac{-95}{49}\right)$$
>
> The magnitude is greater at X than at Y because $|F_x| > |F_Y|$.
>
> The fact that F at X is negative shows that the net force is towards X. The fact that F at Y is positive shows that the net force is towards Y. Therefore, as the satellite moves from X to Y, there exists a point between the two where the net force is zero, so the magnitude decreases and then increases.

Question 9

C Place it in orbit first before trying to attain the required velocity.

Placing the probe into Earth orbit allows a 'slingshot', which uses some of Earth's momentum as an initial velocity, reducing the fuel load required compared with a launch from Earth's surface directly towards deep space. Because $v_{esc} = \sqrt{\dfrac{2GM}{r}}$, the escape velocity is not dependent on the mass of the probe, so **A** and **B** are incorrect. **D** is incorrect, because increasing the force of the engines would increase the velocity of the probe, but this would take more fuel than **C**.

Question 10

A 4 : 3

$$v = \sqrt{\frac{GM}{r}} \qquad v_A = \sqrt{\frac{GM}{2R}} \qquad v_B = \sqrt{\frac{G2M}{3R}}$$

$$v_B : v_A \text{ is } \sqrt{\frac{G2M}{3R}} : \sqrt{\frac{GM}{2R}}$$

$$\sqrt{\frac{2}{3}} : \sqrt{\frac{1}{2}}$$

$$\frac{2}{3} : \frac{1}{2}$$

$$4 : 3$$

Short-answer solutions

Question 11

$\dfrac{mv^2}{r} = \dfrac{GMm}{r^2}$, where m is the mass of the satellite and M is the mass of the planet it is orbiting.

$$\frac{v^2}{r} = \frac{GM}{r^2}$$

$$v = \sqrt{\frac{GM}{r}}$$

Because m is cancelled out, the velocity of the orbiting satellite is independent of its mass.

- 3 marks: correctly correlates relevant formulae to show the orbiting mass is not included in the relationship
- 2 marks: shows the orbiting mass is not included in the relationship
- 1 mark: attempts to equate the relevant formulae

Question 12

$$\frac{r^3}{T^2} = \frac{GM}{4\pi^2}$$

$$T^2 = \frac{4\pi^2(6779 \times 10^3)^3}{(6.67 \times 10^{-11})(6.0 \times 10^{24})}$$

$$\frac{\left((6371 + 408) \times 10^3\right)^3}{T^2} = \frac{(6.67 \times 10^{-11})(6.0 \times 10^{24})}{4\pi^2}$$

$$T = 5543\,\text{s} = 92.4 \text{ minutes}$$

- 3 marks: correctly calculates the period of orbit
- 2 marks: makes a minor error in substitution into the correct equations
- 1 mark: makes some correct substitution into the correct equation

Question 13

In an orbiting spacecraft, astronauts are seen to float as they orbit, which gives the appearance of weightlessness. As weight is defined by the force due to gravity, this could give the idea that there is no gravity. However, it is possible to calculate the gravitational acceleration at this altitude to be $8.9\,\text{m s}^{-2}$, so the astronauts are not actually weightless. They appear to be weightless because there is no reaction force from the spacecraft because it is falling at the same rate as the astronauts. The student's statement is incorrect because the astronaut still has weight.

- 4 marks: makes a judgement of statement based on balanced arguments for and against
- 3 marks: makes arguments for and against
- 2 marks: makes an argument for or against
- 1 mark: makes a relevant statement

Question 14

$$\frac{mv^2}{r} = \frac{GM_E m}{r_E^2} + \frac{GM_S m}{r_S^2}$$

$$\frac{v^2}{(1.515 \times 10^{11})} = \frac{(6.67 \times 10^{-11})(6.0 \times 10^{24})}{(1.5 \times 10^9)^2} + \frac{(6.67 \times 10^{-11})(2.0 \times 10^{30})}{(1.515 \times 10^{11})^2}$$

$$v = 30\,124\,\text{m s}^{-1}$$

$$= 3.0 \times 10^4\,\text{m s}^{-1}\ \text{(to 2 significant figures)}$$

- 3 marks: determines the velocity
- 2 marks: makes a partially correct substitution
- 1 mark: identifies the correct equation

Question 15

Because $2m_A = m_B$, therefore $r_A = 2r_B$.

According to Kepler's third law:

$$\frac{r_A^3}{T_A^2} = \frac{r_B^3}{T_B^2}$$

$$\frac{8r_B^3}{T_A^2} = \frac{r_B^3}{T_B^2}$$

$$\frac{8}{T_A^2} = \frac{1}{T_B^2}$$

$$\frac{T_B^2}{T_A^2} = \frac{1}{8}$$

Therefore, $T_B : T_A = 1 : 2\sqrt{2}$

Thus $\qquad = \dfrac{8}{2\sqrt{2}} = 2.8$ days

- 3 marks: determines the period of B
- 2 marks: makes a partially correct calculation
- 1 mark: provides relevant information

Question 16 ©NESA 2015 MARKING GUIDELINES SIB Q26

a   Model X assumes the Earth's gravitational field is uniform/unchanging/linear from the surface upwards.

Model Y assumes the gravitational field changes with altitude.

- 2 marks: identifies assumptions made
- 1 mark: provides relevant information about an assumption made

b   Variations in gravitational attraction from Earth's surface to an altitude of 200 km are sufficiently small to ensure the results from the two models are not significantly different.

- 1 mark: provides correct reason

c $F_c = F_g$

$$\frac{mv^2}{r} = \frac{GMm}{d^2}$$

$$v^2 = \frac{GMmr}{d^2 m}$$

$$= \frac{6.67 \times 10^{-11} \times 6 \times 10^{24}}{6.58 \times 10^6}$$

$$= 6.08 \times 10^7$$

$$\therefore v = 7797 \, \text{m s}^{-1}$$

- 3 marks: applies a correct method to calculate velocity. States the correct units in the answer
- 2 marks: shows understanding of the relationship between F_c and F_g and attempts at manipulating relevant formulae to find v, **or** attempts to calculate velocity using Kepler's law
- 1 mark: substitutes into a relevant formula

Test 4: Charged particles, conductors, and electric and magnetic fields

Multiple-choice solutions

Question 1

B $2700 \, \text{V m}^{-1}$

$$E = \frac{V}{d} = \frac{80}{0.03} = 2667 \, \text{V m}^{-1}$$

A common error is forgetting to use SI units, such as in **A** where 30 mm is substituted. **C** and **D** incorrectly have the voltage and distance multiplied.

Question 2

A   the same

The electric field between the plates is uniform, meaning that the force per unit charge is the same at all points.

A common misconception is that an individual plate will be the only factor determining the strength of the field ('The charge is closer to the negative plate so it must be stronger'). But this is incorrect because the force is determined by both plates and all points on the plates, so the same force is applied irrespective of where the charge is in that uniform field.

Question 3

B O to P

The work is determined by the force on the charge and its displacement parallel to the direction of the force according to $W = Fs$. For work to be done on a positive charge, it must be applied in the direction opposing the fields lines. **A**, **C** and **D** have shorter displacements, so less work is done.

Question 4

B 1.09×10^{-21} N down the page.

$F = qvB = (1.6 \times 10^{-19})(3.4 \times 10^2)(20 \times 10^{-6})$
    $= 1.09 \times 10^{-21}$ N

This is determined by the right-hand palm rule (or Fleming's left-hand rule). **A** is incorrect because conventional current is in the opposite direction to the electron's motion. **C** and **D** are incorrect because the prefix $\mu = 10^{-6}$ in $B = 20\,\mu$T has not been considered.

Question 5

D 2.2×10^{-3} m

Because $qvB = \dfrac{mv^2}{r}$, then $r = \dfrac{mv}{qB}$.

Although the electron is shown entering at 40° in the diagram, it is travelling perpendicular to the field lines, so the angle is 90°. Therefore, the radius is 2.2×10^{-3} m. **A** is incorrect because incorrect values have been substituted. **B** is incorrect because sin 40 was included. **C** is incorrect because cos 40 was included.

Question 6

A It will bend downwards with an increasing radius of curvature.

Because this is an electron, the direction of the conventional current will be from right to left. Thus, using the right-hand rule, the electron will experience a force downwards because the magnetic field is into the page. As the electron moves from left to right, the magnetic field weakens. Because r is proportional to $\dfrac{1}{B}$, the radius of curvature will increase from left to right.

Question 7

A The proton would pass through undeflected.

The force on the particle from the electric field is in the opposite direction to the force from the magnetic field, so $Eq = qvB$. Thus $v = \dfrac{E}{B}$. Because this is independent of charge and mass, the proton will also pass through undeflected. Both **B** and **C** suggest that a decrease in mass and/or charge would affect the particle's trajectory, but this would only be the case if only one of the fields were present. **D** is incorrect because there is no force acting from left to right.

Question 8

B Velocity, v

$$qvB = \frac{mv^2}{r}$$

$$qB = \frac{mv}{r}$$

but $v = 2\pi rf$

$$qB = \frac{m(2\pi rf)}{r}$$

$$f = \frac{qB}{2m\pi}$$

Thus, f is independent of v.

Question 9

A

Because the field from the plates shown is downward, this will apply a force on the electron in the upward direction. The other field is going into the page and thus will apply a force out of the page.

From the perspective of the electron, this means the force will be up and to the right.

B is incorrect because the student has incorrectly determined the force from the vertical field. **D** is incorrect because the student has incorrectly determined the force from the field into the page.

The most likely incorrect response is **C**. This is from an assumption that the arrows represent the direction of force. This is true for a proton, but not an electron.

Question 10

D It will move a greater distance horizontally, with a smaller radius of turn.

There are two components to the charge's motion: horizontal and vertical. The magnetic field has no effect on its horizontal motion, but the effect of the magnetic field on the vertical component of its velocity will cause it to move in a circular path. The net result will be a spiral path. If the angle is decreased, then the horizontal component of the velocity will increase whereas the vertical component of velocity will decrease. The charge will continue to move horizontally but will travel further.

Because the vertical velocity decreases, according to $qvB = \frac{mv^2}{r}$, a decrease in velocity will cause a reduction in the radius.

A is incorrect because a decrease in the angle will increase the horizontal component of the velocity and therefore the charge will travel further in the same time. **C** is incorrect for the same reason.

B is incorrect because a larger radius is caused by a larger velocity, but the vertical component of the velocity decreases as the angle decreases.

Short-answer solutions

Question 11

Scenario	Electric field	Magnetic field
The charge is stationary.	The charge will accelerate in the direction of the field lines.	The charge will remain stationary.
The charge enters the field at a constant velocity in the direction of the field lines.	The charge will accelerate in the direction of the field lines.	The charge will remain travelling at a constant velocity.
The charge enters the field at a constant velocity perpendicular to the field lines.	The charge will move in a parabolic path.	The charge will move in a circular path.

- 4 marks: correctly describes all scenarios
- 2–3 marks: correctly describes most scenarios
- 1 mark: correctly describes two scenarios

Question 12

a

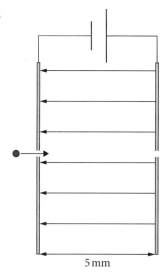

5 mm

- 1 mark: correctly draws the electric field lines between the plates

b $V = Ed = (3000\,\text{V m}^{-1})(5.0 \times 10^{-3}\,\text{m}) = 15\,\text{V}$

The most common error students make is to forget to use SI units for the distance between the plates.

- 2 marks: makes a correct calculation using the correct formula
- 1 mark: attempts a calculation using the correct formula

c
$$W = Vq = \Delta K = \frac{1}{2}m(v^2 - u^2)$$

$$(15)(1.6 \times 10^{-19}) = \frac{1}{2}(9.1 \times 10^{-31})\left(v^2 - (1.0 \times 10^7)^2\right)$$

$$v = 1.0 \times 10^7\,\text{m s}^{-1}$$

Alternatively, students could use $v^2 = u^2 + 2as$, where $a = \dfrac{qE}{m}$.

- 3 marks: correctly calculates the velocity
- 2 marks: makes a correct calculation using a correct formula
- 1 mark: attempts a calculation using the correct formula

Question 13

In the example of the slide, a particle with mass (the child) must increase their potential energy by moving against a gravitational field; for example, to climb the steps to the top of the slide. A charge in an electric field also must move against the field to increase its potential energy. However, this is only true for a positively charged object, because a negatively charged object would increase its potential energy if it moved in the same direction as the electric field. In both cases, when released, the object (the child with mass, or the charge) would accelerate, thereby converting the potential energy to kinetic energy.

Although there are differences, it is a useful model to understand the behaviour of objects in fields.

- 4 marks: identifies one similarity **or** difference between the two situations, and makes an assessment
- 3 marks: identifies one similarity **and** difference between the two situations
- 2 marks: identifies one similarity **or** difference between the two situations
- 1 mark: provides relevant information

Question 14

In a mass spectrometer, a sample is injected and is ionised by the electron beam. The ions from the sample are then accelerated by electric fields to increase their speed. As they enter the magnetic field, they will experience a centripetal force, given by $qvB = \dfrac{mv^2}{r}$.

Rearranging this to $m = \dfrac{qBr}{v}$ shows that the radius curvature is proportional to its mass. The lighter the ions, the more their path curves, i.e. a smaller radius. The detector will record the impact of the ions, separated by their respective curved paths, and thus determine their mass.

- 4 marks: makes a thorough analysis of the function of the mass spectrometer
- 3 marks: makes a sound analysis of the function of the mass spectrometer
- 2 marks: provides a function of the mass spectrometer
  or
  identifies a physical principle in its operation
- 1 mark: provides relevant information

Question 15 ©NESA 2017 MARKING GUIDELINES SIB Q30

The magnetic field will cause the proton to move in a circular path, initially moving out of the page, continuing in an anti-clockwise direction as viewed from the right. At the same time, the electric field will cause the proton to move to the left with increasing speed. The resultant motion of the proton is the vector sum of these two components, which will look like a helix that is getting stretched out towards the left. This helix will be decreasing in radius as the proton loses energy as it radiates electromagnetic radiation as it is an accelerating charge.

Answers may include a labelled diagram.

- 4 marks: provides an analysis of the motion of the proton
- 3 marks: describes the path and/or direction of the proton
- 2 marks: provides some characteristics of the path and/or direction of the proton
- 1 mark: provides some relevant information

Question 16

The electron will undergo projectile motion as it enters the field. Because the electric field is downwards, the electron will experience an upwards force, moving a maximum of 4.0 mm vertically.

$$F = Eq = \frac{Vq}{d} = \frac{(100)(1.6 \times 10^{-19})}{(5 \times 10^{-3})} = 3.2 \times 10^{-15}\,\text{N}$$

Vertically:

$$a = \frac{F}{m} = \frac{(3.2 \times 10^{-15})}{(9.1 \times 10^{-31})} = 3.5 \times 10^{15}\,\text{m}\,\text{s}^{-2}$$

$$u = 0\,\text{m}\,\text{s}^{-1}$$

$$t = ?$$

$$s_y = 4.0 \times 10^{-3}\,\text{m}$$
$$s_y = \frac{1}{2}at^2$$
$$4.0 \times 10^{-3} = \frac{1}{2}(3.5 \times 10^{15})t^2$$
$$t = 1.51 \times 10^{-9}\,\text{s}$$

Horizontally:

$$s_x = vt = (1 \times 10^7)(1.51 \times 10^{-9}) = 1.51 \times 10^{-2}\,\text{m or 15.1 mm (to 2 significant figures)}$$

- 4 marks: correctly determines the range
- 3 marks: attempts calculations to determine the range
- 2 marks: provides a correct calculation
- 1 mark: provides relevant information

Test 5: The motor effect

Multiple-choice solutions

Question 1

C Into the page

The magnetic field is acting downward and the current through the rod is going from left to right, so using the right-hand palm rule, the force will be into the page.

A and **B** are incorrect because it is not possible for a force to be applied parallel to the magnetic field.

D is incorrect and is a result of using the incorrect hand rule (or using the hand rule incorrectly).

Question 2

C The force on the wire remains the same.

The wire is perpendicular to the magnetic field so increasing or decreasing the angle will have no effect on the angle between the wire and the magnetic field lines. Therefore, the force will remain constant.

A and **B** assume incorrectly that the angle shown is the angle θ in $F = BIl\sin\theta$. However, this is the angle between the magnetic field lines and the wire, and this is 90° regardless of θ.

D is incorrect and would only be correct if the wire were moving in the magnetic field. However, the wire is stationary so an EMF is not induced.

Question 3

A 4.9 A left to right

$F_B = F_g$, so $BIl = mg$

$(0.2)(I)(0.5) = (0.05)(9.8)$

Therefore, $I = 4.9$ A

Because the rod is falling downward, the forces from B must act upwards to allow it to fall at constant velocity. Using the right-hand palm rule, this means current must be from left to right.

B is incorrect because the direction is incorrect. **C** and **D** are incorrect because SI units have not been used or have been converted to incorrectly.

Question 4

C Decrease the current

$ma = BIl - mg$

$a = \dfrac{BIl}{m} - g$

Initially the rod is accelerating upward, which means $\dfrac{BIl}{m} > g$

Therefore, to ensure $a = 0$, $\dfrac{BIl}{m}$ has to decrease.

All options except **C** increase the acceleration. Therefore, only decreasing the current will result in $a = 0\,\mathrm{m\,s^{-2}}$.

Question 5

B 1.7 N left

Because $V = IR$

$$I = \frac{V}{R} = \frac{12}{3.0} = 4\,\text{A}$$

$F = BIl$

$= 1.4(4)(0.3)$

$= 1.7$ N

Using the right-hand palm rule, the wire will move to the left.

A is incorrect because voltage has been substituted for current. **C** is incorrect because voltage has been substituted for current and the direction is wrong. **D** is incorrect because the direction is wrong.

Question 6

B 0.05 N

$F = BIl\sin\theta$

$= (1)(2)(0.05)\sin(30) = 0.05$ N

C is incorrect because cos 30 or sin 60 has being used. **D** is incorrect because $\sin\theta$ has been forgotten.

Question 7

B 6.5 A right to left

$$\frac{F}{l} = \frac{\mu_0}{2\pi} \frac{I_1 I_2}{r}$$

$$1.3 \times 10^{-4} = \frac{\mu_0}{2\pi} \frac{2.0 I_2}{0.02}$$

$$I_2 = 6.5\,\text{A}$$

To produce repulsion, the current must be flowing in the opposite direction – from right to left.

A has an incorrect direction. **C** and **D** are incorrect because they have assumed that the current is the same magnitude as in the other wire.

Question 8

A 1.1 T

$$I = \frac{V}{R} = \frac{10}{6.25} = 1.6\ \text{A}$$

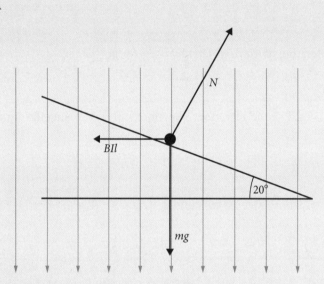

The net force = 0 if the rod moves at constant speed.

$$mg\sin\theta = BIl\cos\theta$$

$$\frac{\sin\theta}{\cos\theta} = \tan\theta = \frac{BIl}{mg}$$

This is consistent with the vector diagram shown below.

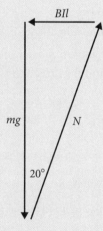

Therefore

$$\tan 20 = \frac{B(1.6)(0.02)}{(0.01)(9.8)}$$

$$B = \frac{(0.01)(9.8)\tan 20}{(1.6)(0.02)} = 1.11\,\text{T}$$

B is incorrect as a result of assuming $mg = BIl$. **C** is incorrect as a result of substituting voltage for current. **D** is incorrect as a result of substituting $\sin\theta$ into the equation.

Question 9

C Axis of rotation: WX; direction of movement: R into the page

Only QP and RQ experience forces because they have components that are perpendicular to the magnetic field. PR is parallel to the field. QP will move out of the page using the right-hand palm rule. RQ will move into the page using the right-hand palm rule. Thus, the axis of rotation will be WX, and R will move into the page.

A is incorrect probably from using the right-hand palm rule incorrectly to determine the force on the loop.

C and **D** have the wrong axis of rotation and do not recognise that there will be no force on PR because the wire is parallel the magnetic field.

Question 10

D AC supply with decreased coil resistance

Decreasing the coil resistance increases the current and thus increases the motor effect force. The greater the force, the greater the movement, and that creates a wave with greater amplitude and therefore louder sound.

A and **B** are incorrect because the current needs to be alternating to make the cone move in and out to produce a sound wave. A direct current would only move the cone in or out. Therefore, no sound would be produced.

Although **D** has alternating current, increasing the resistance decreases the current and thus decreases the force generated.

Short-answer solutions

Question 11   ©NESA   2013 MARKING GUIDELINES SIB Q25

a To the left **or** towards the conductor P

> • 1 mark: identifies correct direction

b $\dfrac{F}{l} = \dfrac{kI_1I_2}{d}$

Force P on Q Force R on Q

$F = \dfrac{2 \times 10^{-7} \times 6 \times 2}{5 \times 10^{-3}}$ $F = \dfrac{2 \times 10^{-7} \times 2 \times 2}{2.5 \times 10^{-3}}$

$F = 4 \times 10^{-4}$ N $F = 3.2 \times 10^{-4}$ N

Total force $= 4.8 \times 10^{-4}$ N $+ 3.2 \times 10^{-4}$ N
$\qquad\qquad = 8 \times 10^{-4}$ N

> • 3 marks: demonstrates correct process to calculate the force experienced by Q
> • 2 marks: demonstrates logical process to calculate force P or force R on Q
> • 1 mark: partial substitution into a relevant equation

Question 12

When the power is switched on, the two rods will carry the same current but in opposite directions. As a result, they will repel each other.

This is because each current-carrying wire generates a cylindrical magnetic field. Each wire will therefore be in the magnetic field from the other wire, and so will experience a force due to the motor effect.

For example, the top wire will generate a magnetic field that will be into the page, which the bottom wire experiences. Therefore, using the right-hand palm rule, the bottom wire will experience a force downwards. Similarly, the top wire will experience a force upwards, consistent with Newton's third law.

- 3 marks: explains the motion of the rod using the motor effect
- 2 marks: attempts to explain the motion of the two rods
- 1 mark: identifies the direction of motion

Question 13

a Because the relationship between force and distance is an inverse relationship, we need to plot F versus $1/d$ to allow us to calculate μ_0 from the graph.

Force ($\times 10^{-5}$ N)	Distance, d (m)	$1/d$ (m)
8.0	0.01	100
4.0	0.02	50
2.7	0.03	33
2.0	0.04	25
1.6	0.05	20

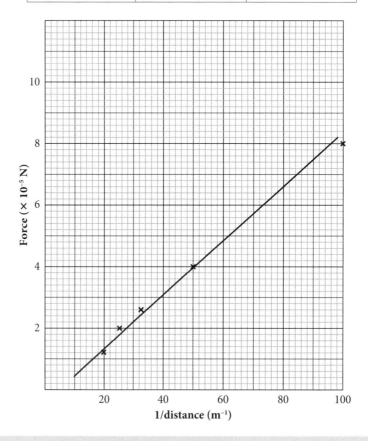

- 3 marks: includes scales, accurate points and a line of best fit
- 2 marks: correctly plots an inverse relationship with minor error
- 1 mark: uses all five data pairs but plots an incorrect graph

b

$$\frac{F}{l} = \frac{\mu_0}{2\pi} \frac{I_1 I_2}{r}$$

$$\text{gradient} = \frac{\frac{F}{1}}{r} = Fr = \frac{\mu_0 I_1 I_2 l}{2\pi}$$

$$\text{gradient} = \frac{y_2 - y_1}{x_2 - x_1} = \frac{8 \times 10^{-5} - 1 \times 10^{-5}}{96 - 16} = 8.8 \times 10^{-7}$$

$$8.8 \times 10^{-7} = \frac{\mu_0 (2)(2)(2)}{2\pi}$$

$$\mu_0 = 2.17\pi \times 10^{-7} \, \text{N}\,\text{A}^{-2}$$

- 3 marks: makes a correct calculation that uses the gradient of the graph
- 2 marks: makes a correct calculation
- 1 mark: provides relevant information

c The value determined of $2.17\pi \times 10^{-7} \, \text{N}\,\text{A}^{-2}$ is about half the value on the data sheet. The graph does not pass through the origin but has an intercept, which suggests a systematic error, because the value is not accurate.

One possible reason for this is that the measuring device, such as systematic errors from incorrect ammeter calibration.

- 3 marks: describes a systematic error with an example
- 2 marks: describes a systematic error
- 1 mark: identifies that it is a systematic error

Question 14

a Because the scale shows a non-zero positive value when the power is turned on, the wire on the scales must experience a force due to the rod above. To do this, the current must be from right to left.

- 1 mark: identifies the correct direction

b $F = mg$

$$= 2.45 \times 10^{-6} \, (9.8)$$

$$= 2.4 \times 10^{-5} \, \text{N}$$

- 1 mark: correctly determines the force

c

$$F = \frac{\mu_0}{2\pi} \frac{I_1 I_2}{r} l$$

$$0.000024 = \frac{\mu_0}{2\pi} \frac{(1.2)I_2}{0.01} \times 0.5$$

$$I_2 = 2 \, \text{A}$$

- 2 marks: correctly determines the current
- 1 mark: correctly substitutes into a relevant formula

Question 15

a $F = BIl = (0.5)(1.5)(0.05) = 0.038\,\text{N}$

> - 2 marks: calculates the correct value for force
> - 1 mark: makes a partially correct substitution into a correct equation

b   Two forces are applied perpendicular to each other.

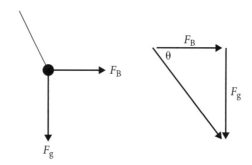

$$\tan\theta = \frac{F_g}{F_B} = \frac{mg}{BIl} = \frac{(0.01)(9.8)}{0.038} = 2.58$$

$\theta = 68.9° = 69°$ (to 2 significant figures)

> - 3 marks: determines the angle
> - 2 marks: attempts to add forces vectorially
> - 1 mark: provides some relevant information

Test 6: Electromagnetic induction

Multiple-choice solutions

Question 1

C   Increase B

> Because flux (ϕ) = $BA\cos\theta$, increasing B will increase flux.
>
> **A** is incorrect because decreasing A will decrease the number of flux lines, decreasing flux. **B** is incorrect because increasing the angle will cause $\cos\theta$ to decrease to zero, thereby decreasing flux. **D** is incorrect because decreasing the magnetic field strength decreases the number of flux lines in the same area.

Question 2

C $8.5 \times 10^{-4}\,\text{Wb}$

> Because $\phi = BA\cos\theta = (0.3)(\pi)(0.03)^2\cos(0) = 8.5 \times 10^{-4}\,\text{Wb}$
>
> **A** is incorrect because the values have not been converted to SI units. **B** is incorrect because π has been forgotten. **D** is incorrect because r has not been squared.

Question 3

D 480 V

$\dfrac{n_2}{n_1} = \dfrac{V_2}{V_1} > 1$, so this is a step-up transformer.

$\dfrac{10}{5} = \dfrac{V}{240}$; therefore, $V = 480$ V

B is incorrect because it has been incorrectly identified as a step-down transformer. This is the most common mistake students make. **C** is incorrect because it assumes there is no change in the voltage. **A** is incorrect because it assumes it is a step-down transformer and divides the voltage by 4.

Question 4

B Wind both coils onto the same nail.

Increasing the flux linkage means increasing the number of magnetic field lines generated by the primary coil that pass through the secondary coil. This can be achieved by using a magnetic material such as iron to 'channel' the field lines between the coils. Therefore, having an iron core (such as a nail) connecting the two loops increases flux linkage. This has the effect of reducing energy loss and this makes it more efficient.

A is incorrect because increasing the length of the coil for the same number of turns will decrease the magnetic field strength. It does not improve the flux linkage.

C is incorrect because increasing the number of turns in both coils while keeping the ratio the same will not significantly affect the efficiency of the transformer. If anything, an increase in the length of wire used will increase resistance and therefore increase energy loss.

D is incorrect because replacing the nails with copper increases the production of small eddy currents and thus decreases efficiency.

Question 5

B

In this case, eddy currents are generated and form parallel to the coil windings. The insulating material disrupts this.

A is incorrect because there is no insulating material.

C is incorrect because eddy currents can still form in the iron sheets.

D is incorrect because there is too much insulation and the rods do not allow for flux linkage with a secondary coil.

Question 6

B A: clockwise; B: anticlockwise

This is an application of Lenz's law. The magnet will fall with a reduced acceleration and some of the loss in gravitational energy will lead to the production of electrical energy in the eddy currents. This is consistent with the law of conservation of energy.

At A the south pole is moving away. Therefore, a north pole must be produced by an eddy current to oppose the motion. Using the right-hand grip rule, this is produced by a clockwise eddy current.

At B the north pole is approaching. Therefore, a north pole must be produced by an eddy current to oppose the motion. This is produced by an anticlockwise eddy current.

For all other options, either one or both directions of eddy current would cause the magnet to accelerate, which increases the rate of change of flux and therefore increases the size of the eddy currents. This would be a violation of the law of conservation of energy.

Question 7

D Greater magnetic flux: Q; greater magnetic flux density: P

Magnetic flux refers to the lines of magnetic force, or field lines. Because there are nine in Q and only four in P, Q has the greatest amount of flux. Magnetic flux density refers to how close the lines are to each other. The density of lines is greater in P.

Question 8

B The current is at a maximum and is in a clockwise direction.

In the position shown, flux is at a minimum. However, the rate of change of flux is at a maximum and therefore an EMF is induced, leading to a current. The right edge of the loop is moving into the page and the left edge is moving out of the page. As the flux threading the loop is increasing, Lenz's law predicts that the induced current should produce flux lines threading to the left within the coil. By the right-hand grip rule, the induced current should flow clockwise in the coil.

A and **C** are incorrect as per the answer above.

D is incorrect. A magnetic field is generated. Because at this point there is a clockwise direction of the current, using the right-hand grip rule shows that the magnetic field goes into the page.

Question 9

C The copper repels the magnet and then attracts it in a continued oscillating motion.

As the copper approaches the magnet, eddy currents are induced that will oppose the change, and as the magnet is free to move, it will be pushed away. As the magnet moves away the copper is then attracted to it, and so on. This is consistent with Lenz's law.

Question 10

B

There must be a change for an EMF to be induced. This occurs at b and d, and will be shown by peaks on the graph. There is no change, so no EMF, at a, c and e. As positive is defined as clockwise, the initial peak should be positive (up) on the graph and the second peak should be negative (down).

Short-answer solutions

Question 11

a $\phi = BA = (0.5)(0.10)^2 = 5.0 \times 10^{-3} \, \text{Wb}$

A common error would be to use incorrect SI unit.

- 2 marks: determines the flux
- 1 mark: makes a partially correct substitution into a correct formula

b The EMF in the initial position is zero so the change in EMF is:

$$\text{EMF} = \frac{\Delta BA\cos\theta}{\Delta t} = \frac{(5\times 10^{-3} - (0.5)(0.1)^2\cos 45)}{0.5} = 3\times 10^{-3}\,\text{V}$$

Because the right side of the wire is starting to move from right to left across the field lines, according to the left-hand palm rule, the EMF will be moving clockwise initially.

Alternatively, as the loop rotates, the flux threading the loop reduces. According to Lenz's law, the induced current will flow to increase the flux threading the loop. Using the right-hand grip rule, this occurs when the induced current is clockwise.

- 3 marks: determines the EMF generated
- 2 marks: determines the magnitude of the EMF generated
- 1 mark: provides relevant information

Question 12

a The velocity is less when the magnet enters the coil than when it leaves. Therefore, the rate of change of flux is less, leading to a lower EMF and a lower current.

- 2 marks: explains why the amplitudes are different
- 1 mark: identifies that the magnet is moving slower as it enters the coil

b

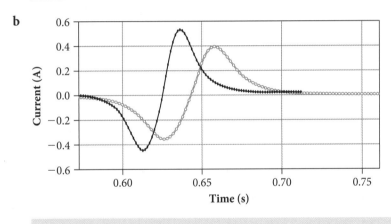

- 2 marks: shows increased amplitude **and** shorter time frame
- 1 mark: shows increased amplitude **or** shorter time frame

c If the magnet is reversed, the trace would be inverted. This is because a change in the direction of the magnetic field would lead to a change in direction of the rate of change of flux and thus a different direction of EMF and current.

- 2 marks: explains why trace is reversed
- 1 mark: identifies that the trace will be reversed

Question 13

Transformers work because of the rate of change of flux, generated by the primary coil and the EMF induced in the secondary coil as it experiences the rate of change of flux.

The efficiency is a comparison of the power input against the power output. 100% efficiency means there is no energy loss, because $(VI)_{in} = (VI)_{out}$. This is affected by flux linkage – the amount of flux generated by the primary coil and thus passing across the secondary coil.

In set-up 1, there is very little flux linkage because of the separation between the two coils. Very little change in flux is experienced in the secondary coil and thus it has zero efficiency.

In set-up 2, the two coils are closer and thus a greater amount of the flux generated is passes over the secondary coil, increasing the efficiency.

In set-up 3, there is increased flux linkage because the secondary coil is inside the primary coil.

The addition of an iron core in set-up 4 concentrates the magnetic field lines, thereby increasing flux linkage further and thus increasing the efficiency to 100%.

- 5 marks: provides a comprehensive analysis of flux linkage **and** refers to multiple aspects of the observations
- 4 marks: provides a thorough analysis of flux linkage **and** efficiency, **and** refers to an aspect of the observations
- 3 marks: provides a sound analysis of flux linkage **or** efficiency, **and** refers to an aspect of the observations
- 2 marks: identifies efficiency **or** flux linkage, **or** outlines an aspect of the experimental methodology
- 1 mark: provides relevant information

Question 14

Power stations can produce electricity at relatively safe voltages but because they are so far from urban centres, the transmission of this electricity would incur huge power losses from the resistance of the transmission lines.

The solution is to use step-up transformers. This results in a large increase in voltage, but it also reduces the current, thus reducing power loss as $P_{loss} = I^2 R$.

$P = VI$, therefore, $I = \dfrac{100 \times 10^6}{5000} = 20\,000\,\text{A}$

$$\frac{V_p}{V_s} = \frac{I_s}{I_p}$$

$$\frac{5000}{130000} = \frac{I_s}{20000}$$

$$I_S = 769\,\text{A}$$

$P_{loss} = I^2 R = (729^2)(40)$

$2.37 \times 10^7\,\text{W}$

In this case, when it reaches customers, the voltage is too high for safe use, and so step-down transformers are used to reduce the voltage to 240 V, a level safe to use within households.

- 4 marks: discusses the role of a step-up transformer **and** the role of a step-down transformer with mathematical analysis
- 3 marks: discusses the role of a step-up transformer **and** the role of a step-down transformer
- 2 marks: discusses the role of a step-up transformer **or** the role of a step-down transformer
- 1 mark: provides relevant information

Question 15 ©NESA 2017 MARKING GUIDELINES SIB Q27

a B (flux density) is given by $\dfrac{\text{flux}}{\text{area}}$ $\left(B = \dfrac{\Phi}{\pi r^2} \right)$

$$B_{\text{max}} = \dfrac{0.6}{\pi \times (0.3)^2}$$
$$= 2.1\,\text{T}$$

- 2 marks: applies the correct process to calculate the maximum magnetic field strength
- 1 mark: provides some relevant information

b Voltage (EMF) = time rate of change of flux. From the graph, this is a maximum when the steepest gradient occurs, i.e. $t = 10-12\,\text{s}$.

$$\text{EMF} = \dfrac{-d\Phi}{dt} = -\left(\dfrac{-0.6}{2} \right) = 0.3\ \text{V}$$

Terminal P is positive and terminal Q is negative (consistent with Lenz's law).

- 3 marks: applies correct process to calculate the maximum voltage and identifies correct polarity of the terminals (P and Q)
- 2 marks: applies correct process to calculate the maximum voltage **or** shows some relevant calculations and identifies the polarity of the terminal(s)
- 1 mark: provides some relevant information

Test 7: Applications of the motor effect

Multiple-choice solutions

Question 1

A 1 matches to 4, 2 matches to 3

1 is an AC generator because it has slip rings and will produce a sinusoidal output. 2 is a DC generator because it has a split ring commutator and will produce a sinusoidal output, but the voltage will always be in one direction because the split ring commutator ensures the voltage in the external circuit remains in the one direction. 3 shows a rectified DC output. 4 shows an AC output. 5 shows constant voltage so is from a battery.

Question 2

A The stator produces a rotating magnetic field.

The rotating magnetic field will cause a change in flux. Eddy currents are produced in the squirrel cage. This results in a force being applied to the squirrel cage and it doesn't rotate.

B is incorrect because there are no brushes in an induction motor. **C** is incorrect because the squirrel cage is not connected to an external power supply and is free to rotate. **D** is incorrect because the rate of rotation is determined by the frequency of the rotating magnetic field that the stator provides. As a result, its rotational speed is variable.

Question 3

A X: back EMF at maximum; Y: load applied

When the drill is turned on, the current rapidly increases and the drill starts to turn because of the motor effect. But as it turns, back EMF reduces the net voltage, the current drops to a minimum and thus the back EMF is at a maximum at X. At Y a load is applied, slowing the rate of turn of the drill, which reduces the rate of change of flux. Therefore, this suddenly reduces the back EMF, increasing the current.

Question 4

A Clockwise

The current is moving in a clockwise direction as seen from the observer. The magnetic field is going from right to left. As a result, the left-hand side of the coil will experience a force upwards. The split-ring commutator will change the direction of the current so that the left-hand side of the coil will always experience an upwards force and the coil will turn in a clockwise direction.

B is incorrect and could have been chosen if the magnetic field direction was incorrect. **C** is incorrect and could have been chosen if the student did not understand the function of a split-ring commutator. **D** is incorrect and could have been chosen if the student mistakenly thought that the current was in the same direction through both sides of the loop.

Question 5

C Current: out of the page; magnetic field: right to left

According to the symbol of the power supply, the current is from Q to P and therefore is out of the page. This set-up requires two opposing torques, so PQ must experience a force downwards to balance the force from the mass at RS. This means that the magnetic field, according to the right-hand rule, must be moving from right to left.

Question 6

B 13.1 A

The torque on both sides must equal. Thus:
$$\tau_{net} = (mg)(r_1) + (BIl)(r_2) = 0$$
$$(5 \times 10^{-4})(9.8)(0.2) = (5 \times 10^{-3})(I)(0.1)(0.15)$$
$$I = 13.1 A$$

A is incorrect because the student has not used correct SI units. **C** is incorrect because the student has not used correct SI units and has forgotten the different radii. **D** is incorrect because the student has forgotten the different radii.

Question 7

A 0.2 N m

$$\tau = nBIA$$
$$= 10(0.2)\left(\frac{10}{3}\right)(0.3 \times 0.1)$$
$$= 0.2 \text{ N m}$$

B is incorrect because the number of turns has not been included. **C** is incorrect because 10 V has mistakenly been used for the current. **D** is incorrect because the area has not been converted to SI units.

Question 8

A Experiment part: disc; AC induction motor part: squirrel cage

In an AC induction motor, a changing magnetic field causes eddy currents to be produced in the rotor. As a result, the rotor will start to turn in the same direction as the changing magnetic field according to Lenz's law.

In the experiment, the rotating magnetic field is demonstrated by moving a magnet. The disc has eddy currents generated in it, which represents the squirrel cage. In an induction motor, it is the stator that produces the rotating magnetic field.

B is incorrect. The magnet is rotating in the experiment, but in an induction motor the magnets form part of the stator, which does not rotate. **C** is incorrect because the disc is moving whereas the stator does not. **D** is incorrect because the disc, not the magnet, corresponds to the squirrel cage.

Question 9

B Conservation of energy sees the loss in kinetic energy result in the production of heat energy.

The production of eddy currents, according to Faraday's law and Lenz's law, causes the metal to slightly heat up because of the resistance in the disc. The disc then slows down, which results in a loss of velocity and thus a loss of kinetic energy. According to the law of conservation of energy, kinetic energy has transformed to heat energy.

A is incorrect because it does not explain why heat is produced by eddy currents. **C** is incorrect because there is no physical contact between the disc and the magnet. **D** is incorrect because, although it is true that the magnetic field generates an opposing force, this does not explain why heat is generated.

Question 10

C $\theta = 60°; \omega = 90°$

The angle θ is the angle between the magnetic field lines and surface vector of the motor. This means the angle is 60°, complementary to the 30° shown in the diagram.

Therefore, for torque, where $\tau = nBIA\sin\theta$, the angle must be 60°.

The angle ω is the angle between the magnetic field lines and the current-carrying wire. This is 90° throughout the rotation of the loop. So for the force from the magnetic field, $F = BIl\sin\omega$, the angle must be 90°.

Although many students will recognise that the wire is perpendicular ($\omega = 90°$) to the magnetic field lines, a common assumption is that the 30° angle is the angle for $\tau = nBIA\sin\theta$, thus suggesting **B** as a possible, but incorrect, response.

Short-answer solutions

Question 11

1. The wires of the coil are not connected to the commutator. As a result no motor effect will be applied to the coil.

2. The winding of the coil changes direction as it goes to the other side of the axle (first clockwise and then anticlockwise). The result is that there will be opposing torques applied to either side of the coil, which results in a net torque of zero.

3. The orientation of the coil is 90° out from the split-ring commutator. In a DC motor, the split-ring commutator switches the current when the torque of the motor is at a minimum. This occurs when the copper brushes momentarily loses contact with the commutators. The result is a torque that maintains its direction.

In the coil's current position, the net torque is zero with both sides experiencing opposing torque because of the windings (maximum torque is achieved when it moves 90° from this position). This means that the current position of the split-ring commutator is 90° out of phase with the coils.

The result is that this motor will not work.

A fourth issue is that there is no current from an external source and no indication of where it is to be connected.

- 3 marks: correctly identifies three mistakes in the construction of the motor
- 2 marks: correctly identifies two mistakes in the construction of the motor
- 1 mark: correctly identifies one mistake in the construction of the motor

Question 12

Back EMF is a result of the motor coil rotating in a magnetic field, resulting in an induced EMF according to Lenz's law. The greater the rate of rotation, the greater the back EMF.

As the rotational speed increases, the EMF increases, the net torque decreases and the rotational speed increases less rapidly.

The amount of back EMF will always be less than the supply EMF, assuming a load on the motor (if there was no load at all, maximum back EMF = supply EMF). The result is a net torque of zero and thus no acceleration. The motor will turn at a constant rate.

For the motor to slow down, the back EMF would have to be greater than the supply EMF, which cannot occur.

Therefore, it is not possible for the back EMF to cause the motor to stop.

- 3 marks: explains why back EMF will not cause the motor to stop spinning
- 2 marks: outlines the effects of back EMF
- 1 mark: provides relevant information

Question 13

The trace shows that the torque changes over time according to $\tau = nBIA \sin\theta$.

At $t = 0$, the torque is zero. This means that the angle between the magnetic field and the surface vector of the loop, θ, is 0°. It turns 180° in 0.5 s, with maximum torque at 90°. Then at 0.5 s, the split-ring commutator reverses the current. This results in the torque remaining in the positive direction. It then completes the loop in the remaining 0.5 s. Therefore, it does one complete revolution in 1 s.

- 3 marks: explains both the magnitude and the direction of the torque in relation to the graph
- 2 marks: describes what the torque does **or** explains why the current remains positive at 0.5 s
- 1 mark: identifies that the torque remains positive **or** identifies that the current changes direction midway

Question 14

The Hair Raiser is in freefall from a height of 50 m; this means it is accelerating at 9.8 m s^{-2}. This is the slope of the line, gradient $= \dfrac{v^2}{s} = 2a$. Freefall can be seen on the graph from 0 m to 25 m. After this, there is a sudden decrease in speed to 10 m s^{-1}.

This decrease in speed can be achieved through magnetic braking. The magnet is turned on at this point and a piece of metal that is attached to the carriage will pass through it, creating eddy currents. This process will always produce an EMF that produces a magnetic field to oppose the motion, and therefore it will slow the motion very suddenly.

In this situation we can see that it causes an initial deceleration of approximately -45 m s^{-2} as determined by the slope from $d = 25$ m to $d = 28$ m. The carriage will then fall more slowly for the remainder of the ride. The graph shows the changing acceleration for the remainder of the fall.

- 4 marks: explains the complete motion of the ride with reference to the graph
- 3 marks: explains the motion without reference to the graph
- 2 marks: describes the motion with reference to the graph
- 1 mark: provides some relevant information

Question 15

If the length of wire is L, then the maximum number of turns $n = \dfrac{L}{2\pi r}$.

The area is πr^2.

Therefore, $\tau = nBIA = BI\dfrac{L}{2\pi r}\pi r^2 = \dfrac{BILr}{2}$

Because $\dfrac{BIL}{2}$ is constant, r is proportional to torque and therefore doubling the radius results in the torque being doubled.

- 3 marks: derives the correct relationship between the torque and the length of wire
- 2 marks: derives a relationship between torque and length
- 1 mark: provides some relevant information

Test 8: Electromagnetic spectrum

Multiple-choice solutions

Question 1

B unifying the laws of electricity and magnetism

By combining the existing laws of electrical interactions and magnetism – Gauss's law, Faraday's law and Ampere's law – Maxwell was able to establish that they are linked, and that electromagnetism behaved as waves travelling at the speed of light.

A is incorrect because, although Maxwell theorised the speed of electromagnetic radiation to be equal to c, he did not do experiments to measure c. **C** is incorrect because Maxwell did not determine the speed of light (scientists before him did). **D** is incorrect because, although Maxwell showed the existence of electromagnetic radiation, he did not 'prove' it, because that would have required experimental evidence to support his model.

Question 2

D X: incandescent bulb; Y: discharge tube; Z: star

X shows the full range of colours and therefore is a continuous spectrum. Y shows only two bands, which means only specific wavelengths are present. This means it is an emission spectrum and therefore is consistent with the discharge tube. Z has specific wavelengths missing, meaning it is an absorption spectrum. A star produces an absorption spectrum.

Question 3

D Mg

If the pattern of lines in the spectrum for each element is mirrored in the spectra of the star, that element is present in the star's atmosphere. The spectrum of magnesium is the only one that is not matched in the pattern in the star's spectrum.

Question 4

C $\dfrac{4d\omega}{\theta}$

The angle between positions 1 and 2 is half of θ.

$\omega = \dfrac{\theta}{2t}$

Therefore, $t = \dfrac{\theta}{2\omega}$

$c = \dfrac{2d}{t} = \dfrac{2d}{\left(\dfrac{\theta}{2\omega}\right)} = \dfrac{4d\omega}{\theta}$

Question 5

D Surface temperature

The surface temperature can only be determined by the relative strength of absorption lines compared with other absorption lines as seen in spectral classes.

A can be determined because the position of the line is determined by the chemicals found in the star's atmosphere. **B** can be determined because the rotation of a star causes a slight redshift on one side of the star and a slight blue shift on the other side, as those two sides move away from and towards the observer respectively. This results in a thickening of the absorption lines. **C** can be determined because as a star moves away, the spectral lines in a spectrum analysis shift towards the red end of the spectrum, i.e. towards longer wavelengths.

Short-answer solutions

Question 6

Source X is a continuous spectrum, and thus all frequencies are present in the spectrum.

Source Y is an absorption spectrum; the light emitted by the source must pass through a medium (such as a gas) that absorbs specific frequencies of light, resulting in the black lines where the specific frequencies are missing.

- 3 marks: explains why the spectra are different
- 2 marks: identifes both spectra **or** identifies one spectrum and explains why it occurs
- 1 mark: identifies one spectrum

Question 7

Historical method	Current method
Armand Fizeau's method	Laser light split and directed through different path lengths
Incandescent light source	Laser light source
The beam is shone at the mirror through the rotating wheel. The speed of the rotation is adjusted until the light is uninterrupted.	The beam is split and directed to the detector. The difference in time is because of the different distances travelled.
Need to know the speed of the wheel and the distance from the mirror to calculate the speed of light.	Need to know the difference in path distance and measure the time difference with an oscilloscope.

- 3 marks: compares methods of determining the speed of light
- 2 marks: outlines methods to calculate the speed of light
- 1 mark: gives characteristics of a method

Question 8

Before Maxwell's work, the concepts of electricity and magnetism were seen as separate domains. Maxwell unified the two by combining a set of equations that describe electric and magnetic fields (Gauss's law) and equations that showed how electric fields and magnetic fields can influence each other (Faraday's law and Ampere's law). The result was that he showed that electric and magnetic fields interact, producing electromagnetic waves (a type of transverse wave) that propagate at the speed of light, c.

This work also suggested that there are other forms of electromagnetic radiation that travel at c. Evidence for this includes the subsequent work of Hertz, who was able to identify radio waves, which also propagate at the speed of light, by measuring their wavelength and frequency.

Another piece of evidence is Essen's determination of the speed of light using microwave cavities, finding the wavelength and frequency of the microwaves to establish the speed of light.

Students do not need to remember the names of the laws mentioned; they are included here for completeness.

- 5 marks: demonstrates a thorough understanding of Maxwell's work, relating to two relevant examples of evidence
- 4 marks: demonstrates a sound understanding of Maxwell's work relating to two relevant examples of evidence or demonstrates a thorough understanding of Maxwell's work relating to one relevant example of evidence
- 3 marks: demonstrates a sound understanding of Maxwell's work **and** provides one relevant example of evidence
- 2 marks: provides one relevant example of evidence **or** demonstrates a sound understanding of Maxwell's work
- 1 mark: provides relevant information

Question 9

The spectrum of the star shows that there is hydrogen in its atmosphere. This can be determined because the emission lines for hydrogen have the same pattern. The core of the star is a black body. It emits all wavelengths, producing a continuous spectrum. When an element is present in the star's atmosphere, certain wavelengths of light are absorbed and are then reradiated in all directions, so that less light of those wavelengths reaches the viewer. This results in dark lines in the star's spectrum.

The pattern of lines is also redshifted. This occurs when the star is moving away from the viewer. According to the Doppler effect, wavelengths of light emitted from a star moving away from the viewer are lengthened and therefore appear closer to the red end of the spectrum.

- 4 marks: explains how two characteristics of the star can be deduced from the spectra provided
- 3 marks: explains one characteristic of the star and describes the other
- 2 marks: correctly links two characteristics of the star with features of the spectra provided
- 1 mark: provides some relevant information

Test 9: Light: wave model

Multiple-choice solutions

Question 1

B Diffraction

Diffraction is the change in direction that occurs when a wave encounters an obstacle or aperture. **A** is incorrect because a minimum of two wave sources is needed for interference to occur. **C** is incorrect because, even though in reality reflection would occur at the wall, it is not shown and thus this image is does not represent reflection. **D** is incorrect because refraction requires the transmission of a wave as it passes through a change in the medium.

Question 2

C 522 nm

$$\tan \theta \approx \sin \theta$$

$$\frac{m\lambda}{d} = \frac{x}{L}$$

$$\lambda = \frac{dx}{L}$$

$$= \frac{(2 \times 10^{-3})(3 \times 10^{-4})}{1.15}$$

$$= 5.22 \times 10^{-7} \text{ m}$$

$$\tan \theta = \sin \theta$$

A and **B** are incorrect because they use the incorrect SI unit for w. **D** is incorrect because it uses the wrong data.

Question 3

D 23 m

$$L = \frac{dy}{m\lambda} = \frac{(0.30 \times 10^{-3})(5.0 \times 10^{-2})}{(1)(640 \times 10^{-9})} = 23.4 \text{ m}$$

A is incorrect because SI units have not been used correctly. **B** and **C** are incorrect because the equation has been rearranged incorrectly.

Question 4

C 0.025 m

$\frac{y}{L}$ must remain constant, so if L is halved, y must be halved.

Question 5

C Newton: light travels as particles; Huygens: light travels as longitudinal waves

Newton thought that light travels as particles or corpuscles, as he referred to them. Huygens thought that light travels as waves in a similar way to sound moving through space, and therefore was a longitudinal wave.

Question 6

D 38%

As the light passes through the first filter, the intensity is halved. When it passes through the second filter, the intensity is reduced to factor of $\cos^2 30$, according to Malus' law. This results in a reduction to 75%. Therefore, as a percentage, the light that will exit the filters is $50 \times 75 = 37.5\%$ of the original.

A is incorrect because this has only calculated the drop in intensity through the first filter. **B** is incorrect because the percentage of light through the filter has been calculated using only $\cos 30$, not $\cos^2 30$, and the first filter has not been considered. **C** is incorrect because the percentage of light through the filter has been calculated using only $\cos^2 30$, without considering the first filter.

Question 7

A 14

Because $d \sin\theta = m\lambda$, then $m = \dfrac{d \sin\theta}{\lambda}$.

The maximum number of maxima occurs when the angle approaches $90°$. Therefore, if we calculate using $\sin 90 = 1$, $m = \dfrac{10 \times 10^{-6}}{680 \times 10^{-9}} = 14.7$

But because m is discrete, the value must be one under $90°$, and is therefore rounded down to 14.

B is incorrect because the value has been rounded up, which would mean that the angle is greater than $90°$, which is not possible. **C** is incorrect because m can only be an integer value. **D** is incorrect because the correct SI units have not been used.

Question 8

C 35°

The second filter is referred to as the analyser, and Malus' law states $I = I_{0\text{max}} \cos^2\theta$.

However, I_{in} is $\dfrac{I_0}{2}$ because the first polariser reduces the intensity by a half.

Therefore

$$\frac{I_0}{2}\cos^2\theta = \frac{I_0}{3}$$

$$\cos^2\theta = \frac{2}{3}$$

$$\theta = \cos^{-1}\left(\sqrt{\frac{2}{3}}\right) = 35.3°$$

A is incorrect and is a result of using $\cos\theta = \dfrac{1}{3}$. **B** is incorrect and is a result of using $\cos^2\theta = \dfrac{1}{3}$.

D is incorrect and is a result of using $\cos^2\theta = \dfrac{1}{6}$.

Question 9

A I, III

> Both Newton and Huygens were able to explain refraction. Newton believed, using the corpuscular model, that corpuscles of light would be attracted to the material in refraction and thus speed up. However, Huygens, using his wave model, predicted that the waves would slow down in the medium using the idea of the path of least distance.
>
> Neither Newton nor Huygens observed polarisation. However, neither could adequately explain why it occurred. Later, polarisation could be explained by using Huygens' wave model, but for this the wave needs to be modelled as a transverse wave.

Question 10

C

> $d \sin\theta = m\lambda$; however, for small angles $\sin\theta \approx \tan\theta = \dfrac{y}{L}$
>
> Therefore, $\dfrac{dy}{L} = m\lambda$
>
> Rearranging, $L = \dfrac{dy}{m\lambda}$
>
> Because $\dfrac{dy}{m}$ is constant, there is an inverse relationship between λ and L.
>
> **A** is incorrect and shows a direct relationship. **B** is incorrect and shows a direct relationship with a negative gradient. **D** is incorrect because it suggests that L is independent of wavelength.

Short-answer solutions

Question 11

a

Who	Newton	Huygens
The model	Light travels as particles, or corpuscles – in essence, particles made up light.	Light produces wavelets of light, making light a longitudinal wave.
Observation	Refraction – Newton passed white light through a prism, which separated it into a spectrum, and then recombined it with another prism. Newton attributed this to different-coloured corpuscles having different sizes.	Refraction could be explained by drawing wavelets and showing geometrically why waves bend when they pass through a medium.

Alternative answers could use reflection or other appropriate observations as the example.

- 4 marks: discusses how each model was developed
- 3 marks: discusses how a model was developed
- 2 marks: identifies the models that Newton and Huygens had
- 1 mark: provides relevant information

b Although Newton's corpuscle model sought to explain refraction, his model suggested that the corpuscles speed up as they enter a dense medium. This was later shown to be incorrect. His model could also not explain polarisation.

Huygens' model was limited in that it could not explain why light travelled in straight lines. It also required a medium for light to travel through, so could not explain why light can travel through a vacuum.

- 2 marks: correctly identifies at least one limitation for each model
- 1 mark: identifies one limitation

Question 12

a There appears to be an angle change of approximately 180°, with the minimum occurring at 90° and 270°. This suggests a cosine relationship.

> - 2 marks: relates the angle to the intensity with reference to the data in the graph
> - 1 mark: relates the variables

b To produce a straight line, the student should plot $\dfrac{I}{I_{max}}$ (the dependent variable) on the vertical axis and $\cos^2\theta$ (the independent variable) on the horizontal axis. This mirrors the relationship in the Malus' law equation.

> - 2 marks: correctly justifies the variables needed to produce a straight line
> - 1 mark: identifies the variables needed to produce straight line

Question 13

White light consists of a range of wavelengths. Each wavelength has different positions for maximum and minimum, determined by $d\sin\theta = m\lambda$.

For example, violet light, with a smaller wavelength, will be diffracted less and will therefore appear closer to the central maximum. Red light, with a longer wavelength, will be diffracted more and will therefore appear further away from the central maximum. The result is that there will be a continuous spectrum from short to long wavelengths as you move away from the central maximum.

> - 4 marks: mathematically relates the diffraction to the pattern and states that longer wavelengths are diffracted a greater amount
> - 3 marks: relates the diffraction to the pattern and states that longer wavelengths are diffracted a greater amount
> - 2 marks: outlines how diffraction contributes to the pattern
> - 1 mark: provides some relevant information

Question 14

$m\lambda = d\sin\theta$, where $\sin\theta \approx \tan\theta = \dfrac{x}{L}$

$$\dfrac{m\lambda}{d} = \dfrac{x}{L}$$

$$\dfrac{2(486 \times 10^{-9})}{2.0 \times 10^{-3}} = \dfrac{x_1}{0.5}$$

$$x_1 = 2.43 \times 10^{-4}$$

$$\dfrac{2(656 \times 10^{-9})}{2.0 \times 10^{-3}} = \dfrac{x_2}{0.5}$$

$$x_2 = 3.28 \times 10^{-4}$$

separation $= x_2 - x_1 = 3.28 \times 10^{-4} - 2.43 \times 10^{-4} = 8.5 \times 10^{-5}\,\text{m}$

> - 4 marks: correctly calculates the separation between the red and green lines at $m = 2$
> - 3 marks: calculates the separation between the red and green lines at $m = 2$
> - 2 marks: attempts to calculate the x value for each wavelength
> - 1 mark: makes a substitution into a relevant formula

Question 15

In the first instance, the angle between the two filters is 90°. According to Malus' law, that means zero transmission because cos 90 = 0. The third filter is placed between the other two at an angle of 45°. This means a transmission of 50% between the first filter and the filter at 45° (because $\frac{I}{I_0} = \cos^2 45 = 0.5$). The light will then be reduced by a further 50% when it passes between the filter at 45° and the final filter. This means a net reduction to 25%. This means there is still some light passing through all three filters, as can be seen in the image.

- 3 marks: explains how the light is affected by each filter, resulting in the transmission of light
- 2 marks: discusses the transmission of light
- 1 mark: provides some relevant information

Question 16

The positions of the visible emission lines can be determined by $d \sin\theta = m\lambda$ if light is acting as a wave, with wavelengths. However, diffraction occurs with all forms of electromagnetic radiation, not just visible light. Hence the model can be used to predict the positions of UV and IR emission lines. Appropriate technology can be used to verify their existence.

- 4 marks: demonstrates a thorough understanding of how the wave model of light, as evidenced by diffraction, can be used to verify the existence of emission lines in the ultraviolet or infrared range of the spectrum
- 3 marks: demonstrates a sound understanding of how the wave model of light, as evidenced by diffraction, can be used to verify the existence of emission lines in the ultraviolet or infrared range of the spectrum
- 2 marks: identifies that the model can be used to detect either infrared or visible or ultraviolet light
- 1 mark: provides relevant information

Test 10: Light: quantum model

Multiple-choice solutions

Question 1

B The wavelength with the highest spectral irradiance

According to Wien's displacement law, $T = \frac{b}{\lambda}$, the temperature is inversely proportional to the peak wavelength on the black body curve, in this case approximately 520 nm.

A is incorrect because the area under the graph represents the total intensity emitted by the black body. **C** is incorrect because the peak spectral irradiance just tells us the intensity of the peak wavelength. **D** is incorrect because most black bodies emit a similar range of wavelengths, although their intensities vary.

Question 2

B The leaves don't move.

The experiment works because UV radiation has a high energy according to $E = hf$, which causes electrons to be ejected from the metal, and as a result the leaves move together. Red light has photons with a lower frequency and therefore a lower energy. As a result, the electrons do not gain enough energy to be ejected from the metal and therefore the gold leaves don't move.

All other responses incorrectly suggest that red light has enough energy for photoelectrons to be emitted.

Question 3

C Temperature decreases; intensity decreases; wavelength increases

A black body curve shows intensity for every wavelength for a given temperature (spectral irradiance). As the temperature increases, the wavelength with the peak spectral irradiance decreases. Therefore, decreasing the temperature will increase this wavelength, and the intensities at all wavelengths will decrease.

A is incorrect because it states that as temperature increases, irradiance decreases. **B** is incorrect because it states that as temperature increases, wavelength increases. **D** is incorrect because it states that as temperature decreases, spectral irradiance increases.

Question 4

D 12×10^{14} Hz

The threshold frequency is the minimum frequency at which incident photons have enough energy for photoelectrons to be emitted. For beryllium, this photon energy is $5\,eV$.

$$E = hf$$
$$5(1.6 \times 10^{-19}) = 6.626 \times 10^{-34}(f)$$
$$f = 1.207 \times 10^{15}\,Hz$$
$$= 12 \times 10^{14}\,Hz$$

A is incorrect because it quotes the energy for beryllium in eV. **B** is incorrect because it quotes the energy for caesium. **C** is much too large to be correct.

Question 5

C Current decreases; voltage increases

Because the frequency of the violet light is greater than the frequency of the green light, the photoelectrons being emitted have more energy. This requires a greater stopping voltage.

Because the intensity of the light is reduced, the number of photons is reduced and therefore the number of photoelectrons emitted is also reduced, thus decreasing the current.

Question 6

B 800 nm

The threshold frequency for potassium is 0.375×10^{15} Hz.

$$v = f\lambda$$
$$\lambda = \frac{3 \times 10^8}{0.375 \times 10^{15}}$$
$$= 0.0000008\,m$$
$$= 800\,nm$$

A and **D** are incorrect as a result of misinterpreting the frequency for the wavelength.

C is incorrect as a result of misreading platinum for potassium.

Question 7

C 5×10^{-19} J

$E = hf_0$

Sodium: $E = (6.63 \times 10^{-34})(0.625 \times 10^{15}) = 4.14 \times 10^{-19}$ J

Tungsten: $E = (6.63 \times 10^{-34})(1.375 \times 10^{15}) = 9.11 \times 10^{-19}$ J

The difference $= 5 \times 10^{-19}$ J

A is incorrect because this is the work function for sodium. **B** is incorrect because this is the work function for tungsten. **D** is incorrect because this is the difference between threshold frequencies.

Question 8

B For metal X, the number of photoelectrons emitted would increase, but the maximum kinetic energy would remain unchanged.

Light with a wavelength of 450 nm has an equivalent frequency of 6.7×10^{14} Hz.

According to the graph, this frequency is above the threshold frequency for X, and therefore photoelectrons are emitted.

However, this is below the threshold frequency for Y, and therefore no photoelectrons are emitted.

Increasing the intensity of the light increases the number of photons, but it does not increase their energy and therefore will not change the kinetic energy of the photoelectrons.

A is incorrect because kinetic energy will only increase if there is an increase in the frequency (i.e. a decrease in the wavelength). **C** and **D** are both incorrect because no photoelectrons are emitted for Y.

Question 9

B There are fewer photons with high energy (i.e. short wavelengths).

On a black body curve, the y-axis represents the spectral irradiance, which is the intensity for each given wavelength. Because the intensity is low, the number of photons for that wavelength is less.

A is incorrect because short-wavelength photons have a high frequency and therefore high energy. **C** is incorrect because black bodies absorb all forms of radiation, irrespective of their wavelengths. **D** is incorrect because it does not address the question. Also, interference is a wave phenomenon and thus cannot be used to support a quantum phenomenon.

Question 10

D Voltage will decrease but current will increase.

The voltmeter measures the stopping voltage and therefore can measure the energy of the photoelectrons that are emitted. That energy is determined by $E = hf - \varphi$, where φ is the work function of the metal.

Because the frequency of the incident light is decreased, the energy of the emitted photoelectrons decreases and thus the stopping voltage will decrease.

The current measures the number of photoelectrons that are emitted, which is determined by the intensity of the incident light. If the intensity of the light increases, the number of photoelectrons and therefore the current must increase.

Short-answer solutions

Question 11

In graph A, electron energy versus frequency, it can be seen that the frequency needs to be above a certain value before electrons can be ejected. At this frequency, electrons are ejected with no kinetic energy, and above this frequency, electrons have kinetic energy. This suggests that light energy is delivered in quanta called photons. If the frequency is below the threshold frequency, it will not have enough energy to cause photoelectrons to be emitted even over time – it is not possible to accumulate enough energy from low-frequency light. This is supported by $K = hf - \phi$.

In graph B, the kinetic energy does not change as the intensity changes. This is because the energy of the photoelectrons is determined by the frequency of the incident light photons according to $E = hf$.

Graph C shows that changing the frequency of the incident light does not change the number of photoelectrons emitted (i.e. the current that is measured).

Graph D shows that as intensity is increased, the current increases. This indicates that more electrons will be ejected with higher intensity incident light. However, it does not suggest they will have more energy.

The quantum nature of light is demonstrated by the fact that a minimum amount of energy (i.e. a minimum frequency) is needed for photoelectrons to be emitted. Only a change in frequency will affect the energy of photoelectrons emitted. A greater intensity of light (i.e. more photons) will not cause photoelectrons to be emitted if it is not of the correct frequency.

- 5–6 marks: explains how the data provided suggests quantum nature of light
- 4 marks: describes how the data provided suggests quantum nature of light
- 3 marks: describes some data in relation to quantum nature
- 2 marks: describes some of the data
- 1 mark: provides relevant information

Question 12

Planck understood that the existing mathematical model for the relationship between energy and temperature could not explain the shape of the black body radiation curve.

By assuming that energy is quantised, where $E = hf$, he reformulated the mathematical equation known at that time. As a result, the new mathematical equation could model the experimental evidence of the black body radiation curve.

- 3 marks: describes Planck's work in determining a mathematical model for the black body
- 2 marks: shows some understanding of Planck's work
- 1 mark: identifies that the mathematical model for black body radiation was not consistent with the experimental curve

Question 13

a

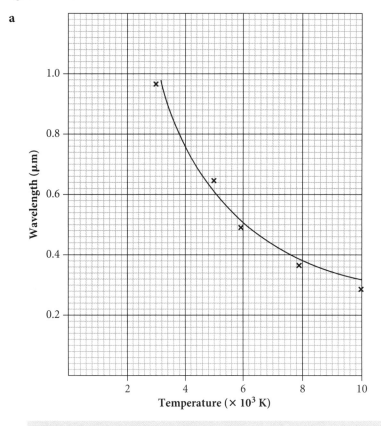

- 3 marks: correctly plots points with correct axes **and** has a trendline
- 2 marks: correctly plots points with correct axes where temperature is on the x-axis and wavelength is on the y-axis **or** correctly plots points with a trendline
- 1 mark: correctly plots points

b

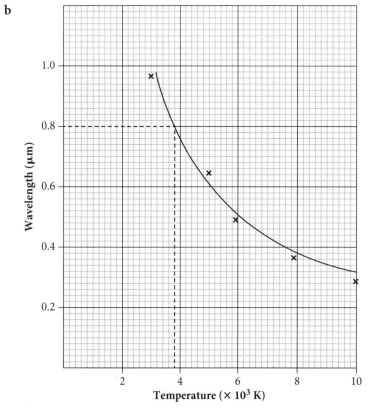

The temperature is 3900 K.

- 2 marks: correctly determines the temperature from the graph
- 1 mark: attempts to use the graph to determine the temperature

c The area underneath the black body curve represents the total intensity of radiation emitted.

As well as the peak wavelength shifting to shorter wavelengths, an increased temperature also results in an increased area. This is shown by the increased value of the intensities in the data.

- 2 marks: outlines that the intensity is related to the area under the graph and increased temperature results in increased intensity and thus area
- 1 mark: identifies that the intensity is related to the area under the graph

Question 14

$$f = \frac{c}{\lambda} = \frac{(3.0 \times 10^8)}{(599 \times 10^{-9})} = 5.0 \times 10^{14}\,\text{Hz}$$

$$E = hf = (6.626 \times 10^{-34})(5.0 \times 10^{14}) = 3.31 \times 10^{-19}\ \text{J} = 2.1\text{eV}$$

Because this is lower than the work function of copper, photoelectrons will not be emitted.

- 4 marks: correctly determines the energy in eV to determine if photoelectrons are emitted
- 3 marks: determines the energy to determine if photoelectrons are emitted
- 2 marks: determines the energy of the photon
- 1 mark: correctly substitutes into an equation

Question 15

a

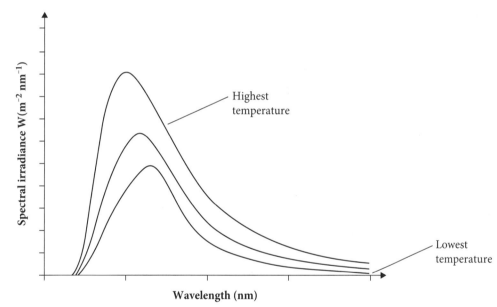

- 3 marks: sketches a series of labelled curves of differing temperatures depicting characteristics of peak shifting to smaller wavelengths and higher peaks as temperature increases; there are no graph overlaps and each graph with a higher temperature has a larger area beneath it
- 2 marks: sketches a series of curves of differing temperatures and depicts some characteristics that show the changes
- 1 mark: provides some relevant information

b The peak of the curve is found and then the corresponding wavelength. This is then substituted into the equation for Wien's law, $T = \frac{b}{\lambda}$, to determine the temperature.

- 2 marks: correctly outlines the method, including the mathematical formula that is needed
- 1 mark: provides some relevant information

Test 11: Light and special relativity

Multiple-choice solutions

Question 1

B 4.5×10^{17} J

$E = mc^2 = (5.0)(3.0 \times 10^8)^2 = 4.5 \times 10^{17}$ J

A is incorrect because the value of c has not been squared. **C** is incorrect because the value of c has not been squared and correct SI units have not been used. **D** is incorrect because correct SI units have not been used.

Question 2

D   do nothing because there is no way to tell from within your reference frame.

Because you are in an inertial frame of reference, the laws of physics are no different from those acting in a stationary frame of reference. Therefore, there is no experiment you could perform that would differentiate between the two inertial frames of reference.

The only way you could tell if you were moving would be if you were accelerating and therefore were in a non-inertial frame of reference.

A is incorrect because the accelerometer is travelling at the same constant speed as you are, so it would show no deviation. **B** is incorrect because the ball would fall the same way whether you were not moving or moving at constant velocity, so this would not tell you anything. **C** is incorrect because you would always see your reflection, no matter how fast you are travelling.

Question 3

C   Clock A starts before clock B.

Light is travelling at the same speed in both directions according to the observer. However, the rear of the train, clock A, is moving towards the observer – the distance the light has to travel is reduced. Clock B, at the end of the train, is moving away from the observer and therefore the distance the light has to travel is increased. Therefore, clock A starts first and then clock B.

A is incorrect because the path length the light travels from the two clocks to the person at Y is different. **B** is incorrect because the direction has been misinterpreted.

Question 4

C   on the spaceship measuring the time taken for light to travel from the front to the back of the spaceship.

To measure the rest-frame length, an experiment must be performed to measure the time a pulse of light takes to move a set distance inside the frame of reference that is moving; that is, the observer must be inside the spacecraft, measuring the spacecraft.

A is incorrect because the distance used is outside the frame of reference of the spacecraft. **B** and **D** are incorrect because the measurement is made on the planet.

Question 5

B A space station in orbit

In reality, the question is asking for an example of a non-inertial frame of reference. This means the frame of reference is one that is accelerating. Because the space station is in orbit, it is experiencing a centripetal acceleration and thus is non-inertial.

A is incorrect because although this capsule is falling and is under the influence of gravity, the fact that it is travelling at terminal velocity means it is no longer accelerating.

C is incorrect because the train is travelling at $0.5c$, a constant velocity, and therefore is not accelerating.

D is incorrect because if the lift is travelling at a constant rate of change of displacement, it is travelling at a constant velocity, and therefore it is not accelerating.

Question 6

D The frequency will be larger than expected, but the wavelength will be the shorter than expected.

If a light source moves towards an observer, according to the Doppler effect its light will be blue shifted and hence its observed frequency will increase. The light therefore becomes blue shifted.

However, because the velocity of light must remain constant $(c = f\lambda)$, an increasing frequency will result in a decrease in wavelength.

A and **B** are incorrect. If only the frequency changes, the velocity of the wave $(c = f\lambda)$ would change as well. This is not possible because c is constant.

C is incorrect because it has frequency getting lower and wavelength longer, which is the opposite of what is observed.

Question 7

B $1.23 \times 10^{20}\,\text{Hz}$

A positron has an identical mass to an electron. The combined mass of the positron and electron converts to energy when they annihilate according to $E = mc^2$.

However, this energy is shared between two photons.

$E = (m_e + m_p)c^2 = 2(m_e)c^2$

$E = hf$

$$f = \frac{\frac{1}{2}\left(2(m_e)c^2\right)}{h} = \frac{(9.109 \times 10^{-31})(3.0 \times 10^8)^2}{6.626 \times 10^{-34}} = 1.23 \times 10^{20}\,\text{Hz}$$

A is incorrect as a result of confusing energy with frequency. **C** is incorrect as a result of not recognising that the energy is shared between photons. **D** is incorrect as a result of confusing energy with frequency and not recognising that the energy is shared between two photons.

Question 8

B

According to the observer, the ball is falling at a constant acceleration. However, the ball is also accelerating to the left. This is because the ball is now moving at a constant velocity relative to the track horizontally but, according to the observer, is accelerating away.

Because it is moving at a constant acceleration in both the x and y directions, the net result is a straight-line motion as seen in **B**.

A is incorrect as it suggests that the ball is continuing to accelerate with the observer. **C** is incorrect as it suggests that the ball is travelling at a greater velocity than the observer. **D** is incorrect as it suggests that the ball is moving away but, relative to the observer, is moving at a constant velocity. This would suggest, incorrectly, that the ball is still accelerating horizontally relative to the track.

Question 9

B Time: 10.6 ms; length: 943 km

The stationary observer measures the time $t = \dfrac{d}{v} = \dfrac{1.0 \times 10^6}{\left(\dfrac{3.0 \times 10^8}{3}\right)} = 10\,\text{ms}$. This is the proper time.

1000 km is measured in the stationary observer's frame of reference and is the proper length.

Therefore, the measurement on the spacecraft will be dilated in time and contracted in length.

A is incorrect because although it correctly identifies time dilation, the length should not also be dilated. **C** is incorrect because it has incorrectly determined the time (suggesting it contracts) and the length (suggesting it dilates). **D** is incorrect because it has incorrectly determined the time.

Question 10

C X sees it take a longer time to reach the sensor than Y.

Both X and Y see the light pulse travelling at speed c. Relative to X, the light pulse must travel half the length of the train. However, Y sees a shorter distance because the back of the train is moving towards them. Therefore, the time it takes for Y is shorter than the time it takes for X.

A is incorrect because the speed of light is always c. **B** is incorrect because X always sees the same distance; however the distance for Y is smaller because the rear of the train is moving towards them. **D** is incorrect because there is always a difference in measured lengths and times for observers in different frames of reference.

Short-answer solutions

Question 11

a Mass defect $= \dfrac{E}{c^2}$

$\qquad\quad = \dfrac{4 \times 10^{-12}}{(3.0 \times 10^8)^2}$

$\qquad\quad = 4.4 \times 10^{-29}\,\text{kg}$

- 1 mark: provides the correct mass defect

b $E = 4.0 \times 10^{-12}\,\text{J}$ for one reaction

Total energy in 1.0 s $= 3.8 \times 10^{26}\,\text{J}$

Therefore, there are $\dfrac{3.8 \times 10^{26}}{4.0 \times 10^{-12}} = 9.5 \times 10^{37}$ reactions each second (to 2 significant figures).

Therefore, mass lost $= (9.5 \times 10^{37}) \times (4.4 \times 10^{-29}\,\text{kg}) = 4.2 \times 10^9\,\text{kg s}^{-1}$

- 2 marks: determines the mass lost
- 1 mark: makes a correct calculation

c $4.0 \times 10^{-12}\,\text{J}$ energy is produced per reaction, so there needs to be $\dfrac{(46 \times 10^6)}{(4.0 \times 10^{-12})} = 1.15 \times 10^{19}$ reactions to produce the same energy

Each reaction has a mass defect of $4.4 \times 10^{-29}\,\text{kg}$.

The combined mass lost is $(4.4 \times 10^{-29}) \times (1.15 \times 10^{19}) = 5.1 \times 10^{-10}\,\text{kg}$

- 2 marks: correctly calculates the mass
- 1 mark: makes a correct calculation

Question 12

a $l = l_0\sqrt{1 - \frac{v^2}{c^2}} = (6300 \times 10^3)\sqrt{1 - \frac{7500^2}{c^2}} = 6299.999\,998 \text{ km}$

> - 2 marks: calculates the contracted length correctly calculates the proper distance
> - 1 mark: identifies it as length contraction

b $t = \dfrac{6299.999\,998 \text{ km}}{7.5 \dfrac{\text{km}}{\text{s}}} = 839.999\,9857 \text{ s}$

The results show that, even at high speeds of a satellite, there is very little relativistic effect.

> - 1 mark: calculates the correct time

Question 13

0.25 g of antimatter will annihilate with exactly 0.25 g matter; therefore, the total mass is 0.5 g

Thus $E = mc^2 = (0.5 \times 10^{-3})(3.0 \times 10^8)^2 = 4.5 \times 10^{13} \text{J} = 45\,000 \text{ GJ}$

5000 tonnes $\rightarrow$ 5000 $\times$ (4.185 GJ) = 20 920 GJ

Therefore, the claim is incorrect. In fact, the antimatter will generate twice as much energy.

Matter and antimatter will only annihilate with the same type of fundamental particle.

For example, an electron will annihilate with a positron (anti-electron) but will not annihilate with an anti-quark.

> - 4 marks: calculates the energy of the matter–antimatter annihilation and the amount of energy from dynamite, and assesses the claim numerically
> - 3 marks: determines the energy of the matter–antimatter annihilation and assesses the claim
> - 2 marks: determines the energy of the matter–antimatter annihilation
> - 1 mark: provides relevant information

Question 14

a Distance = velocity $\times$ time = $(0.98)(3.0 \times 10^8)(1.56 \times 10^{-6})$ = 458 m

> - 1 mark: determines the length a muon would travel in one half-life

b With a rest half-life of 1.56×10^{-6} s, a muon will only travel 458 m.

Due to the random nature of radioactive decay, muons have a range of life spans, both less than and more than the stated half-life. The mean life span is 2.2×10^{-6} s.

Even with this time, they would only travel 460 m on average. This means that very few muons would reach the laboratory.

> - 2 marks: accounts for the low number of muons detected
> - 1 mark: provides relevant information

c   Because the muon is travelling at 0.98c, the distance travelled relative to the muon's frame of reference is significantly shorter, because of length contraction.

$$l_v = l_0 \sqrt{1 - \frac{v^2}{c^2}}$$

$$= 10000\sqrt{1 - (0.98^2)} = 1989\,\text{m}$$

As a result, more muons are detected because there is an increased number that can travel the shorter distance before decaying.

- 2 marks: accounts for the increased number of muons detected
- 1 mark: determines the contracted length

Question 15

a   The graph shows an exponential growth of momentum as the speed approaches c. This is also shown in

$$p_v = \gamma p_0$$

where

$$\gamma = \frac{1}{\sqrt{1 - \frac{v^2}{c^2}}}$$

As $v \to c, \gamma \to \infty$

- 2 marks: describes the relationship between momentum and speed in the graph **and** with an equation
- 1 mark: describes the relationship between momentum and speed in the graph **or** with an equation

b   A car that is crashing will have measurable changes in momentum based on $p = mv$, and no momentum dilation will be observed because the velocity is too slow.

This is verified in the dilation formula:

$$p_v = \gamma p_0$$

where

$$\gamma = \frac{1}{\sqrt{1 - \frac{v^2}{c^2}}}$$

Because $\frac{v^2}{c^2}$ will reduce to zero at low speeds, γ will be equal to 1, so $p_v = p_0.$

- 2 marks: justifies why the speed of a car is too slow to get measurable relativistic effects
- 1 mark: identifies that the speed of a car is too slow

Question 16

Einstein's special theory of relativity (SR) is based on the postulates that the speed of light is constant for all frames of reference and that all inertial frames of references are equivalent. These lead to mathematical models.

Models lead to predictions that can be tested. The predictions in SR include that time measurements are relative, as are length and momentum – two observers in different frames of reference will measure any event as having differences in time, length and/or momentum.

SOLUTIONS **227**

The technology to test this was not available when Einstein proposed SR. However, decades later other scientists were able to test the predictions made by SR. They included the measurement of muons, created in the upper atmosphere. Their rest half-life (t_0) is significantly smaller than their measured half-life when measured from Earth (t_v), because they are travelling with a velocity (v), close to the speed of light (c).

The values measured are consistent with the time dilation formula:

$$t_v = \frac{t_0}{\sqrt{1 - \frac{v^2}{c^2}}}$$

which validates this model.

Measurements made in particle accelerators provide evidence for length contraction and momentum dilation, which are consistent with the mathematical models resulting from SR.

- **4 marks**: demonstrates a thorough understanding of the SR model and how experimental evidence validates the theory
- **3 marks**: outlines how SR was supported by experimental evidence
- **2 marks**: demonstrates some understanding of SR **or** demonstrates some understanding of the scientific model
- **1 mark**: provides relevant information

Test 12: Origins of the elements

Multiple-choice solutions

Question 1

C $4\,^1H \rightarrow {}^4He$

The diagram represents the proton–proton chain and shows a series of events that lead to the formation of a helium nucleus. Although hydrogen nuclei are products in some stages of the process, the net result means that there is an overall loss of four hydrogen nuclei for a gain of one helium nucleus.

A is incorrect. The process shows six hydrogen nuclei being used as reactants, but it also shows two hydrogen nuclei as products, hence only four net hydrogen nuclei. **B** is incorrect because only one helium atom is produced. **D** is incorrect because hydrogen is a reactant but is not included in this equation.

Question 2

C S has a higher surface temperature than R.

This H–R diagram has increasing luminosity as you go up and decreasing temperature as you go from left to right. Therefore, S has a higher surface temperature than R.

A is incorrect. Because S is lower on the scale it has a lower luminosity than Q. **B** is incorrect because R has a lower surface temperature than S and therefore will be more red. **D** is incorrect. P sits on the main sequence, which are stars undergoing hydrogen fusion. R is a red giant, and this is undergoing helium fusion and therefore is older than P.

Question 3

A A blue star that has high luminosity

Spectral class relates to the temperature of stars. A star with a spectral class of B is a hot star, of approximately 20 000 K. Because it is large, the best example is Spica.

B is incorrect. A white dwarf is hot and can have a spectral class of B, but it is very small, and thus its luminosity is very low. The example here is Sirius B.

C is incorrect. Stars that are on the main sequence with low luminosity are cooler in nature. The example shown is Proxima Centauri.

D is incorrect. A cool star will not have a spectral class of B. If it has a high luminosity, it must be very large. A good example is Betelgeuse.

Question 4

C A random selection of stars found on the main sequence could all have the same age.

Stars on the main sequence are those stars that fuse hydrogen. The position is determined by the magnitude and temperature. As a result, stars found on the bottom right of the diagram will have longer lifespans because they are smaller and fuse hydrogen at a lower rate. Stars in the upper left are larger and fuse hydrogen at a higher rate and therefore have shorter life spans. However, life span is not the same as age, so it is possible to find many stars with different lifespans that are the same age.

A is incorrect because a large absolute magnitude means it is dimmer. **B** is incorrect because as a star ages, it stays on the main sequence. When it starts to fuse helium, it will jump to the giant or supergiant phase, which is to the right of the main sequence. **D** is incorrect. Although many stars that are older can be found to the right of the diagram, some can be found elsewhere. For example, a white dwarf can be very old and is found on the left side of the diagram.

Question 5

A Spica is denser than the Sun.

Because Spica is denser, a greater number of fusion reactions are taking place in any given volume. As a result, the rate of reactions is higher, which will result in an increased luminosity and temperature.

B is incorrect. Although Spica is larger than the Sun, this is not the determining factor. For example, red giants are significantly larger than the Sun, but because of the increased surface area, their surface temperatures are actually lower than the Sun.

C is incorrect. Although technically true, this is not the reason for it being more luminous.

D is incorrect because luminosity is about intrinsic brightness and therefore distance plays no part.

Question 6

A 1.9 billion years

The slope is the Hubble constant (H_0).

The inverse of the slope $\left(\dfrac{1}{H_0} \right)$ is $\dfrac{\text{distance}}{\text{velocity}} = \text{time}$

$\dfrac{(2 \times 10^6)(3.086 \times 10^{16})}{10^6} = 6.172 \times 10^{16}$ s. This is 1.9 billion years.

B is incorrect. Although this is the actual estimate for the age of the Universe, it is not based directly on Hubble's data, as Hubble made certain assumptions of distances which are now known to be incorrect.

C and **D** are incorrect because the correct SI units have not been used.

Question 7

A A

> A white dwarf is the stage at the end of the star's life cycle for smaller stars. They are carbon cores. They have still extremely high temperatures, so will sit to the left of the H–R diagram. However, because they are very small, their brightness is very low. Thus, their absolute magnitude is a larger positive number, placing it in the lower part of the H–R diagram.

Question 8

D D

> The Sun is an average star and sits approximately in the middle of the main sequence. If the Sun increased in luminosity, it would move up the H–R *diagram* but maintain its spectral class, meaning it would move neither left nor right. Thus, the best option is **D**. It is likely that the Sun will eventually turn into a red giant at its end stages, will and move to that position on the diagram.

Question 9

A The galaxies are moving away from us.

> The spectral lines have been redshifted. This means that the light source is moving away from us.
>
> **B** is incorrect. If they are moving towards us, that would be shown by a move towards the blue end of the spectrum. **C** and **D** are incorrect because the patterns are identical and thus represent the same element, which is calcium.

Question 10

C The production of matter from energy

> As the Universe expanded, its density decreased. At a certain point, the drop in temperature resulted in energy being converted into the fundamental building blocks of matter. In turn, this increased the rate at which the temperature of the Universe decreased.
>
> **A** is incorrect because the inflationary period happened in the first second of the Universe forming. **B** is incorrect because stars and galaxies formed later as gravity started to be the dominant force in the Universe. **D** is incorrect because although there was a drop in temperature, it was the result of a drop in density, not the other way around.

Short-answer solutions

Question 11

The evidence presented in this question supports the expansion of the Universe.

The first piece of data is the Cosmic Microwave Background (CMB) radiation. It shows the Universe as far back in time as we can see, at the point when the Universe became transparent (the light we see has its wavelengths stretched into the microwave range). The temperature of the Universe at that time was fairly uniform. The only way that it could be so uniform would be if all of the Universe was once together and then spread apart. This supports the expansion of the Universe.

The second piece of data is a graph produced by Hubble, with the speed of galaxies plotted against their distance. The redshift in the spectra of the galaxies showed them to be receding from Earth in all directions. The data can best be explained by the Universe expanding in all directions away from the observer.

The last piece of data is the table of the abundance of gases in the Universe. Hydrogen, which is the simplest form of atomic matter, has the greatest abundance, which is consistent with the process of matter formation in the early Universe.

These three pieces of evidence all support the theory of the expansion of the Universe.

- 5 marks: demonstrates a thorough understanding of the three pieces of information
- 4 marks: demonstrates a sound understanding of the three pieces of information
- 3 marks: demonstrates a sound understanding of two pieces of information
- 2 marks: outlines a piece of evidence using the data provided **or** identifies the three pieces of evidence
- 1 mark: provides relevant information

Question 12

Sample answer

	G2	K1 red giant
Main fusion	p–p chain	Triple alpha
Energy output	1.44 MeV	7.2 MeV
Characteristics	Stable	Stable
Colour	Yellow	Red
Temperature	Approx 5000–6000 K	3000 K
Relative radius	Small	Large

A G2 star is in equilibrium. This means that the energy that is produced by fusion is balanced by the gravity pulling the matter together. A K1 red giant is also in equilibrium, but because the energy from its fusion is so much greater, its radius increases to a much larger value before it is balanced by gravity. As the radius increases, the density of the gas decreases, which in turn decreases the temperature of the surface gas, and it will therefore appear red.

- 5 marks: demonstrates an extensive understanding of how the energy released from fusion reactions defines the characteristics of stars
- 4 marks: demonstrates a thorough understanding of how the energy released from fusion reactions defines the characteristics of stars
- 3 marks: demonstrates a sound understanding of how the energy released from fusion reactions defines the characteristics of stars
- 2 marks: identifies the reaction in relation to the evolutionary stage of the stars
- 1 mark: provides some relevant information

Question 13

The core of a star is considered to be like a black body. It therefore produces a characteristic black body radiation curve. This is a continuous spectrum. The temperature of the star depends on its size, and this can be determined by the peak wavelength of the spectrum.

Because the atmosphere of the star contains cooler gas, this will include gases that are cool enough for atoms to hold electrons. The radiation from the core will excite these electrons. As the electrons fall back down to lower energy levels, they will release energy in particular wavelengths specific to the element. These wavelengths of light are emitted in all directions, so an observer will see less of this wavelength of light. This is an absorption spectrum, and it corresponds to the dips in the curve.

- 3 marks: explains how the star's atmosphere absorbs specific wavelengths to produce a modified black body curve
- 2 marks: outlines how a black body curve is modified because of light absorption
- 1 mark: identifies the curve as a black body curve **or** identifies that some light is being absorbed

Question 14

In the model, the raisins (galaxies) move further apart over time. This is similar to the Big Bang theory because galaxies move further apart over time. The model suggests that as the bread doubles in size, the spacing between the raisins also doubles, suggesting an even increase in all directions. This represents the expansion as described in the Big Bang theory closely. It does not account for observations that show galaxies moving towards each other or moving at a greater speed.

The model shows that the 'Universe', i.e. the bread, expands over time and the galaxies move within this space. Similarly, the Big Bang theory says that the Universe gets bigger. Although the bread itself expands into space, the Universe does not expand into anything. The raisins are embedded in the dough. There is a suggestion that the dough carries the raisins out during expansion, whereas in the Universe, the galaxies are in space.

Although the model does not represent all aspects of the Big Bang, it does accurately represent the main ideas and therefore helps with understanding.

- 4 marks: assesses how well the model reflects the Big Bang
- 3 marks: relates some features of the model to the Big Bang
- 2 marks: describes some features of the model
- 1 mark: provides some relevant information

Question 15

All stars in the main sequence convert hydrogen to helium. Their temperatures will be different, depending on their masses, and therefore they will appear at different positions on the H–R diagram. This will then indicate the luminosity of the star.

The Sun and Spica are both main sequence stars. Spica is much higher on the main sequence curve because it has a greater mass and therefore higher temperature.

The graph incorporating luminosity versus temperature shows that the Sun is a low-mass star and therefore uses the p–p chain as the dominant form of fusion.

Spica has a much greater mass and luminosity and is considered a high-mass star. The dominant form of fusion would be CNO. The reactions in these stars occur more often and therefore the stars have more power, resulting in a higher luminosity. The CNO cycle relies not only on the star's mass but also on the presence of carbon, nitrogen and oxygen.

- 5 marks: explains both reactions and relates them to the H–R diagram
- 4 marks: explains a reaction and relates it to the H–R diagram
- 3 marks: describes the reactions that occur and relates these to the H–R diagram
- 2 marks: identifies reactions that occur in the stars
- 1 mark: provides some relevant information

Question 16

In a main sequence star, the forces within the star are in balance. The two forces that are present are gravity (inward) and the force exerted from the release of energy from fusion (outward).

The amount of mass collected as the star begins to form will determine what size and colour it will be. The larger the mass collected, the greater its gravitational pull. This will cause the gas to heat up and allow fusion to begin. The larger the mass, the more reactions that will occur and therefore the greater outward force from the release of energy from fusion. The star will then grow in size until the two forces balance.

- 3 marks: explains the balance of forces in relation to the size and brightness of a star
- 2 marks: describes features of a stage in stellar evolution
- 1 mark: provides some relevant information

Test 13 Structure of the atom

Multiple-choice solutions

Question 1

C

Geiger and Marsden initially used polonium, an alpha source, to probe gold atoms. The prevailing model at the time was Thomson's plum pudding model, which modelled the atom as a sphere with negatively charged electrons embedded in it. The expectation was that some alpha particles would deflect slightly but most would pass through the gold foil.

B and **D** are incorrect because they incorrectly refer to protons. **A** is incorrect because, although this image shows the actual result Geiger and Marsden obtained and the model that resulted, it was not what was expected.

Question 2

B   travel in straight lines.

The formation of a shadow showed that the cathode ray travelled in straight lines. To show that cathode rays were negatively charged, they would need to have been deflected by an electric or magnetic field, neither of which is shown in the image. Therefore, **A** is incorrect. **C** and **D** are incorrect because cathode rays are not waves but are composed of electrons.

Question 3

A   I only

(I) has electric plates, which allow the user to apply a perpendicular electric field. This would show the negative nature of the charges.

(II) has a paddle wheel. This demonstrates that the charges have momentum.

(III) has a Maltese cross. This shows that the cathode rays travel in straight lines.

Question 4

D   Suspending a negatively charged particle with an electric field going downwards

Robert Millikan suspended negatively charged oil droplets in an electric field in such a way that the force from the electric field balanced the force from gravity. Therefore, the electric field needed to apply an upwards force on the negatively charged particles, so the electric field lines must be downwards.

A and **B** are incorrect because Millikan used negatively charged oil drops. **C** is incorrect because an upward electric field would apply a force on a negatively charged particle in the downward direction, in the *same* direction as gravity.

Question 5

C   Rutherford's model could explain the recoil of alpha particles.

The recoil of alpha particles in the Geiger–Marsden experiment suggested a small heavy nucleus, which could not be explained by Thomson's model.

A is incorrect because although Rutherford's model was later modified by Bohr, whose model could explain spectral lines, Rutherford's model alone could not. **B** is incorrect because in both models, alpha particles could pass through gold foil. **D** is incorrect because although Rutherford's model is referred as a planetary model, this does not explain why it was accepted.

Question 6

D charge-to-mass ratio of the electron.

A and **B** are incorrect because both refer to Millikan's work, not Thomson's. **C** is incorrect because the mass of the electron cannot be determined directly from the density of the oil drop, and oil drops were a feature of Millikan's work.

Question 7

B It could not be deflected by magnetic fields and yet had mass.

The particle that was identified was the neutron, which is neutral in charge. The experiment showed it has mass but no charge.

A is incorrect because its mass is approximately a quarter of an alpha particle. **C** is incorrect because it is neutral. **D** is incorrect because it must have momentum in order to produce knock-on protons.

Question 8

D $q = \dfrac{mgd}{V}$

Because the oil drop fell at a constant rate, the force from the electric field is equal in magnitude to the force due to gravity.

$$F_E = F_g$$

$$Eq = mg$$

$$\frac{V}{d}q = mg$$

$$q = \frac{mgd}{V}$$

Question 9

D 3.8 mT

B is incorrect because correct SI units have not been used. **A** and **C** are incorrect because the equation has been rearranged incorrectly.

$$F_B = F_c$$

$$qvB = \frac{mv^2}{r}$$

$$\frac{q}{m} = \frac{v}{Br}$$

$$1.76 \times 10^{11} = \frac{2.0 \times 10^7}{B(0.030)}$$

$$B = 3.79 \times 10^{-3}\,\text{T}$$

Question 10

D The force from the electric field is stronger than the force from the magnetic field but is neither parabolic nor circular.

The electrons are now experiencing two forces – the force from the electric field (F_E), which is upward, and the force from the magnetic field (F_B), which is initially downward.

Because the beam curves upwards, this means that $F_E > F_B$.

As the electrons start to change direction, the magnitude of F_E remains constant.

The magnitude of F_B changes because it is increasing in velocity, but it is also changing direction.

Only the horizontal component of the velocity contributes to a downward force, and if it is assumed the horizontal velocity does not change, neither does the magnitude of F_B.

Therefore, the net force in the vertical would remain as a constant value: $F_{net} = F_E - F_B$.

This results in an acceleration vertically upwards, and the path would be parabolic.

But this is if the horizontal velocity remains constant.

However, the magnetic field also applies a force horizontally, to the left, against the electron beam.

This increasingly slows the electrons and the path curves more because the magnitude of F_B reduces.

The result is that it will **not** produce a parabolic shape (nor a circular shape, because that requires a constant F_B acting inwards).

A and **C** are incorrect because this would mean the path would curve downward.

B is incorrect because it has been assumed that the horizontal velocity of the beam is not affected.

Short-answer solutions

Question 11

The results from the Geiger–Marsden experiment did not reflect the Thomson model. If the Thomson model was correct, the alpha particles would pass through undeflected, as shown on the left.

The deflection that was actually noted could not be explained by the Thomson model. However, the deflection could be explained if there was a very small positive area at the centre of the atom. This explained the observed spread of the alpha particles, because the positive mass at the centre of the atom would repel the particles and deflect them. Therefore, Rutherford developed his new model to account for the observations that were made.

- 3 marks: makes links from the observations of the experiment to features of the model
- 2 marks: makes an observation that links to the model
- 1 mark: provides some relevant information

Question 12

$$D = \frac{m}{V}$$

$$900 = \frac{m}{\frac{4}{3}\pi(312 \times 10^{-9})^3}$$

$$m = 1.14 \times 10^{-16}\,\text{kg}$$

$$F_g = F_E$$

$$mg = Eq = \frac{V}{d}q$$

$$(1.14 \times 10^{-16})(9.8) = \frac{11.7}{3.0 \times 10^{-3}}q$$

$$q = 2.86 \times 10^{-19}\,\text{C}$$

Allow for a range of radius values from 310 to 315 nm.

- 3 marks: correctly determines the charge of an electron
- 2 marks: attempts to make a calculation of q
- 1 mark: makes a substitution into a relevant formula

Question 13   ©NESA 2014 MARKING GUIDELINES SIB Q28

a

- 3 marks: draws an annotated diagram that clearly shows how Thomson's experiment can be performed
- 2 marks: draws a diagram that shows some features of Thomson's experiment
- 1 mark: shows a basic understanding of Thomson's experiment

b Equating $F = qvB$ and $F = \dfrac{mv^2}{r}$

$$qvB = \frac{mv^2}{r}$$

$$r = \frac{mv}{qB}$$

$$= \frac{(9.109 \times 10^{-31}) \times (1 \times 10^{7})}{(1.602 \times 10^{-19}) \times (9 \times 10^{-4})}$$

$$= 0.063\,\text{m}$$

- 3 marks: shows correct process to calculate the radius of the electron's path
- 2 marks: equates $F = qvB$ and $F = \dfrac{mv^2}{r}$ to find r **or** shows partial substitution into relevant formulae
- 1 mark: shows partial substitution into a relevant formula

Question 14

The student's results show that most of the balls move under the cardboard and continue in a straight line. Only a few balls are either deflected or come back to where they were launched. These are similar results to those found by Geiger and Marsden. In their experiment, alpha particles were fired at gold foil. The ball that is rolled under the cardboard is modelling the alpha particles that were fired at the gold foil. The object under the board represents a nucleus in the gold foil.

In the student's experiment, most balls rolled straight through, modelling that the atom is mainly empty space. Only a few balls were deflected; these represent alpha particles that came close to the gold nuclei. The ball that came back to its launch point represents an alpha particle hitting the nucleus.

- 4 marks: explains the student results seen and relates this to the Geiger–Marsden experiment
- 3 marks: describes the student results seen and relates this to the Geiger–Marsden experiment
- 2 marks: describes the results
- 1 mark: provides some relevant information

Question 15

Rutherford had observed that the mass of the helium nucleus suggested that it was the equivalent of 4 protons, but its charge did not agree with this. The atomic number of helium is 2, which means it has 2 protons. As its mass number is 4, the nucleus was measured to be more massive than could be explained by the presence of 2 protons, hence a new particle was suggested.

- 2 marks: relates Rutherford's observations to the need for the neutron
- 1 mark: makes reference to unexplained observations

Question 16

In 1897, Thomson developed a model of the atom after he discovered that electrons were negatively charged particles. Because the atom was considered neutral, to balance these negative charges there must be some positive charge.

Therefore, his model consisted of a sphere of positive charge with negatively charged electrons embedded into it, like a plum pudding with raisins. It thus became called the plum pudding model.

In 1908, Geiger and Marsden devised an experiment whereby they fired alpha particles at gold foil. Based on the plum pudding model, they expected most of the alpha particles to pass through undeflected.

However, some alpha particles recoiled, providing evidence that was inconsistent with the plum pudding model. This demonstrates scientific methodology: observations were made that were not consistent with the model at that time, which led to a new model being developed.

Rutherford developed a new model of the atom, in which most of the mass of the atom resided in a small, dense nucleus, with electrons in orbit around it. This new model could account for the experimental results of Geiger and Marsden.

Alternative answers can include Rutherford's view of the nucleus with Chadwick's discovery of the neutron. It could also mention the modification of the Rutherford model by Bohr, or the model then developed by de Broglie.

- 7 marks: refers to the work of two relevant scientists and demonstrates a thorough knowledge of how models change based on experimental evidence
- 5–6 marks: describes the work of two relevant scientists **and/or** using examples, demonstrates a sound understanding of the development of a scientific model
- 3–4 marks: outlines the work of a relevant scientist **and/or** using an example, demonstrates a basic understanding of the development of a scientific model
- 2 marks: identifies a relevant scientist **or** identifies a relevant experiment
- 1 mark: provides relevant information

Test 14: Quantum mechanical nature of the atom

Multiple-choice solutions

Question 1

B It provided an explanation for emission spectra.

By stating that electrons could jump between discrete orbits with an emission of a specific energy ($E = hf$), it was able to explain the production of absorption and emission spectra.

A is incorrect because although it accounted for stable orbits, Bohr was not able to explain why they were stable. **C** is incorrect because the model was a mix of both classical and quantum physics. **D** is incorrect because it was Rutherford's model that formed the basis for the Bohr model.

Question 2

D 6.56×10^{-7} m

$$\frac{1}{\lambda} = R\left(\frac{1}{n_f^2} - \frac{1}{n_i^2}\right)$$

$$\frac{1}{\lambda} = 1.097 \times 10^7 \left(\frac{1}{2^2} - \frac{1}{3^2}\right)$$

$$\lambda = 6.56 \times 10^{-7} \text{ m}$$

A is incorrect as a result of not squaring the 'n' values and not finding the reciprocal to find λ.
B is incorrect as a result of not finding the reciprocal to find λ. **C** is incorrect because both n values
have not been squared.

Question 3

C 4.84×10^{-19} J

$$\frac{1}{\lambda} = R\left(\frac{1}{n_f^2} - \frac{1}{n_i^2}\right)$$

$$\frac{1}{\lambda} = 1.097 \times 10^7 \left(\frac{1}{2^2} - \frac{1}{6^2}\right)$$

$$\lambda = 4.102 \times 10^{-7} \text{ m}$$

$$E = hf = \frac{hc}{\lambda} = \frac{(6.626 \times 10^{-34})(3 \times 10^8)}{4.102 \times 10^{-7}} = 4.84 \times 10^{-19} \text{ J}$$

A is incorrect because it quotes the frequency. **B** is incorrect because λ instead of f has been substituted.
D is incorrect because the n values were not squared.

Question 4

B 3.14 nm

When the electron is in the $n = 2$ state, according to de Broglie there are two wavelengths around the
atom in resonance; thus the circumference of the orbit is equal to the wavelength.

$$n\lambda = 2\pi r = \frac{2\pi(9.99 \times 10^{-10})}{2} = 3.14 \times 10^{-9} \text{ m}$$

A is incorrect because it is simply restating the radius. **B** is incorrect because it is the ground state value,
not $n = 2$. **C** is incorrect because it has the wavelength doubled, not halved from ground state.

Question 5

B 1.5×10^{-11} m

$$\lambda = \frac{h}{mv} = \frac{6.626 \times 10^{-34}}{(9.109 \times 10^{-31})(5.0 \times 10^7)} = 1.5 \times 10^{-11} \text{ m}$$

A is incorrect because only mass has been substituted in. **C** is incorrect because an incorrect formula
has been used.

Question 6

A Elements produced unique emission spectra consisting of discrete wavelengths.

B is incorrect because these observations were made much later, in bubble chambers. **C** is incorrect
because this was observed by Geiger and Marsden under the direction of Rutherford. **D** is incorrect
because this was observed by Davisson and Germer, providing the evidence of electron wave behaviour.

Question 7

C IR-B

For $n_{4 \to 3}$

$$\frac{1}{\lambda} = R\left(\frac{1}{n_f^2} - \frac{1}{n_i^2}\right) = 1.097 \times 10^7\left(\frac{1}{3^2} - \frac{1}{4^2}\right) = 5.333 \times 10^5$$

$\lambda = 1.88 \times 10^{-6} = 1.88\,\mu m = 1800\,nm$

Therefore, it is in the IR-B range.

Question 8

A The position of each electron is described as a probability.

B is incorrect because although Schrödinger's model did have a central nucleus, it was Rutherford who established this. **C** is incorrect because although Schrödinger's model did have distinct energy levels, the hypothesis was made by Bohr. **D** is incorrect because although Schrödinger's model did have negative charges, the hypothesis was made by Thomson.

Question 9

C Bohr: energy orbitals; Schrödinger: probability regions

Bohr described the electrons as only being able to exist in specific energy orbitals around the nucleus, whereas Schrödinger described electron positions as probability clouds, where electrons were likely to be found around the nucleus.

A and **B** are incorrect because the regions where electrons are found are better described as probability regions. **D** is incorrect because energy levels are better used to describe Bohr's model.

Question 10

C absorbed or released by the atom is limited to specific values consistent with the Rydberg equation.

Packets of light, with an energy $E = hf$ (according to Bohr) cause an electron to jump from a lower energy orbit to a higher energy orbit. This energy is discrete and is consistent with Rydberg's formula. **A** is incorrect because light is not absorbed gradually – the energy required must be in specific amounts, where $E = hf = \frac{hc}{\lambda}$. **B** is incorrect because although the light emitted is equivalent to the drop in energy, it is not its cause. It is the drop in energy levels that results in a release of energy with a wavelength consistent with Rydberg's formula. **D** is incorrect because the energy released by the atom must be equal to the energy absorbed.

Short-answer solutions

Question 11

a   The minimum amount of energy is when an electron moves from $n = 2$ to $n = 3$.

$$\frac{1}{\lambda} = R\left(\frac{1}{n_f^2} - \frac{1}{n_i^2}\right)$$

$$\frac{1}{\lambda} = 1.097 \times 10^7 \left(\frac{1}{2^2} - \frac{1}{3^2}\right)$$

$$\lambda = 6.56 \times 10^{-7} \text{ m}$$

$$E = hf = \frac{hc}{\lambda} = \frac{(6.626 \times 10^{-34})(3 \times 10^8)}{6.56 \times 10^{-7}} = 3.03 \times 10^{-19} \text{ J}$$

- 2 marks: correctly calculates the energy
- 1 mark: substitutes into a relevant equation

b   The Balmer series comprises the emission lines of hydrogen in the visible wavelengths. These are a series of wavelengths that excite electrons to jump to a higher level and then back down to $n = 2$. It is noted that these are discrete, which suggests that not all wavelengths of light can cause this and therefore the energy is quantised. The Rydberg formula helps us predict which wavelengths can cause this.

- 3 marks: explains emission spectra and its relevance to the quantum nature of light
- 2 marks: describes spectra
- 1 mark: provides some relevant information

Question 12

$$\lambda = \frac{h}{mv}$$

$$6.56 \times 10^{-7} = \frac{(6.626 \times 10^{-34})}{9.109 \times 10^{-31} v}$$

$$v = 1.11 \times 10^3 \text{ m s}^{-1}$$

- 2 marks: correctly calculates the velocity
- 1 mark: substitutes into a relevant equation

Question 13

Rutherford's model of the atom had most of the mass of the atom located in the centre and was positively charged. This came about after the observations of Geiger and Marsden. It explained why the alpha particles that were fired at the gold foil did not behave as expected. Although this explained the observations and improved on the plum pudding model, it did not explain how the electrons remained in stable orbits around the nucleus; according to the model, they would spiral in as they lost energy.

Bohr's model sought to address the stability of the atom by placing the electrons in specific energy levels (orbits) and assumed that the energy for stable orbits was quantised. This model could also explain the spectrum of hydrogen. However, this model also had its limitations. It did not explain other spectra or the Zeeman effect, and it could not explain why orbits were stable without this quantum condition.

- 5 marks: presents features of Rutherford's and Bohr's models of the atom and assesses their limitations
- 4 marks: presents features of Rutherford's and Bohr's models of the atom and mentions their limitations
- 3 marks: describes the Rutherford's and Bohr's models and identifies their limitations
- 2 marks: describes Rutherford's and Bohr's models
- 1 mark: presents some relevant information

Question 14

De Broglie's model of the atom suggested that electrons in their orbits behaved as waves. He suggested that electrons are only found in discrete regions because they are in an orbit that produces a standing wave. In all other orbits, the size of the orbit would cause the electron wave to interact in a destructive interference pattern and therefore fall into another orbit.

The pattern shown in the diagram is that of an interference pattern. As the electrons are reflected from different layers of the nickel crystal, they will either interfere with each other in a destructive or constructive way, depending on the difference in path length (i.e. the size of the crystal spacing).

A diffraction pattern is a characteristic phenomenon of waves. Because this is produced when electrons are shot at the nickel crystal, it supports the idea proposed by de Broglie that electrons are waves.

- 3 marks: describes features of each experiment and relates how observations support de Broglie's model of the atom
- 2 marks: describes a feature of the experiment and outlines a feature of the de Broglie model
- 1 mark: describes features of each experiment

Question 15

Bohr's model of the atom describes electrons as particles and in discrete orbits. However, his model could not explain why those orbits were discrete. His model could also only describe the hydrogen atom.

Schrödinger's atom, building on the idea of electrons acting as waves that was first suggested by de Broglie, described the electron orbits based on probability and therefore electron orbits are discrete probability clouds. Not only did his model explain the orbits, it was also applicable to all elements.

Schrödinger's model is much more complex than Bohr's model. It is therefore appropriate to use Bohr's model to introduce discrete orbits as a starting point in learning about the atom. Bohr's model helps explain the formation of bonds and valency in chemistry. Simplified models can still be used, even though they may be incomplete or inaccurate.

- 4 marks: demonstrates a thorough understanding of Schrödinger's and Bohr's models and relates the benefits of using Bohr's model
- 3 marks: demonstrates a sound understanding of Schrödinger's and Bohr's models and identifies a benefit of using an inferior model
- 2 marks: demonstrates a basic understanding of Schrödinger's and Bohr's models and identifies a benefit of using an inferior model
- 1 mark: demonstrates a basic understanding of Schrödinger's or Bohr's model

Question 16 ©NESA 2013 MARKING GUIDELINES SII Q35e

De Broglie was able to utilise the current understanding of matter to further develop our understanding of the structure of matter. At the time there was an understanding of the nature of light, which was energy, displaying particle-like properties. Also, a model of the atom was established displaying quantised electron energy levels, but this was yet to be explained. De Broglie utilised the ideas about light being a particle and applied this to electrons, stating that it could behave as a wave. By establishing that electrons exist in stationary states that correspond to the wavelength of the electron, and that they could only move between energy levels that correspond to this wavelength, de Broglie was able to provide an explanation for the stability and quantised nature of the electrons in the atom. De Broglie was then able to describe the wave particle duality of matter. In this way de Broglie increased our understanding of the nature and structure of matter.

- 6 marks: shows a thorough understanding of the contributions made by de Broglie **and** describes how de Broglie used existing concepts and ideas to come up with new interpretations, **and** uses clear and scientifically correct language
- 5 marks: shows a good understanding of the contributions made by de Broglie **and** outlines how de Broglie used existing concepts/ideas to come up with new interpretations
- 4 marks: shows a sound understanding of the contributions made by de Broglie **and** identifies existing concepts/ideas de Broglie used to come up with new interpretations in our understanding of the structure of the matter
- 3 marks: shows some understanding of the contributions made by de Broglie
- 2 marks: identifies some contribution(s) made by de Broglie and/or existing concept(s) and/or idea(s)
- 1 mark: identifies a contribution made by de Broglie or existing concept or idea

Test 15 Properties of the nucleus

Multiple-choice solutions

Question 1

B   0.0069

The decay constant, $\lambda = \dfrac{\ln 2}{t_{1/2}}$

The half-life in the graph can be determined when half of the mass remains. Because it starts at 400 g, $t = 100$ years when the mass is 200 g

Therefore, $\lambda = \dfrac{\ln 2}{100} = 0.0069$ per year.

A is incorrect as a result of using $\log 2$ instead of $\ln 2$. **C** is incorrect as it simply quotes the numeral for $\ln 2$. **D** is incorrect as it simply quotes the time passed.

Question 2

B   X: alpha; Y: gamma; Z: beta

Both X and Y are charged particles because they curve in a magnetic field.

Using the right-hand palm rule, the direction of X indicates it must be a positive charge; therefore, it is an alpha particle. It also has a radius two times greater than Z, because it is 4 times heavier and has twice the charge.

Using the right-hand palm rule, Z must be a negative charge; therefore, it is a beta particle.

Because Y is undeflected, it must be a gamma photon.

A is incorrect because it has Y as a beta particle, which would not travel straight in a magnetic field. **C** is incorrect because it has the particles for X and Z back to front. **D** is incorrect because it has Y as an alpha particle, which would not travel straight in a magnetic field.

Question 3

C The binding energy per nucleon is greater for each of the products than the reactants.

During a fission reaction, a large nucleus splits into two smaller nuclei and there is a decrease in mass between the reactant and the products. This mass defect results in an increased energy according to $E = mc^2$ and thus the binding energy per nucleon for the products increases.

A is incorrect. It does occur with any larger nuclei, but for energy to be released it can only be from $Z = 56$ (iron) and above. **B** is incorrect because there is no loss of nucleons during a fission reaction. **D** is incorrect because the combined mass of the products is less than the combined mass of the reactants.

Question 4

B Control rods

Control rods control the rate of fission by absorbing neutrons. They can stop the fission process if they are inserted fully into the core.

A is incorrect because the moderator's role is to slow neutrons. **C** is incorrect because the heat exchanger is responsible for transferring the thermal energy of the core to other materials such as water to produce steam. **D** is incorrect because it is in the core where fission takes place.

Question 5

C Process R: α; product Q: Th-230

Arrows to the left and down show a drop in both the atomic number and the neutron number. This means there is a loss of 2 protons and 2 neutrons. Thus, they represent alpha decay.

Arrows to the right show an increase in atomic number but a decrease in neutron number. This means that a neutron transmutes into a proton. Conservation of charge means that it must also liberate an electron or beta particle.

Because Q has 90 protons and 140 neutrons, its mass number will be 230 and the element must be thorium.

B and **D** have the incorrect process. **A** has incorrectly identified the product Q.

Question 6

B Beta, cobalt

^{56}Ni has 28 protons and 28 neutrons. According to the graph, it will undergo β^+ decay. This is a result of a proton transmuting to a neutron, and thus releasing a β^+ particle or positron. Therefore, the atomic number must decrease to 27, which is cobalt.

A is incorrect. For nickel to turn into copper it will undergo beta decay, but it will be in the form of β^-. **C** and **D** are incorrect. It will not undergo alpha decay, and, if it did, it would become iron.

Question 7

C a gamma source because it can penetrate deeply through most things.

Gamma radiation would be seen to be coming from the pipe, with fractures seen as hot spots, with an increased release of gamma radiation.

A is incorrect. Although beta sources have a short half-life, they cannot penetrate deeply into the soil and thus to the Geiger counter.

B is incorrect because alpha particles do not have enough penetrating power.

D is incorrect for the same reason as **B**.

Question 8

C 9.7 g

$N = N_0 e^{-\lambda t}$ where $\lambda = \dfrac{\ln 2}{t_{\frac{1}{2}}}$

$N = (35)e^{-\frac{\ln 2}{1622}(3000)}$

$\quad = 9.7\,\text{g}$

A and **D** are incorrect because the decay constant, λ, has not been used in the calculation. **B** is incorrect as a result of simply determining the fraction from $\dfrac{1622}{3000}$.

Question 9

A H + H → He

The fusion reaction has the greatest increase in binding energy per nucleon, which means it has the greatest mass defect and thus releases the most energy.

B is also a fusion reaction but has a lower increase in binding energy per nucleon, so is incorrect. **C** is incorrect because it is a fission reaction and so has a smaller increase in binding energy per nucleon. **D** is incorrect because it *requires* energy to fuse iron (Fe).

Question 10

A increases when fused and therefore releases energy.

When the binding energy of an atom increases, it will release energy during the reaction.

Elements that have a smaller atomic number than iron will do this when they undergo fusion, whereas elements that have a greater atomic number than iron can only achieve this through fission.

Short-answer solutions

Question 11

a When there is 12.5% of ^{60}Co remaining, 3 half-lives have passed. This is 15.81 years.

- 1 mark: correctly determines the half-life of ^{60}Co

b $N = N_0 e^{-\lambda t}$ where $\lambda = \dfrac{\ln 2}{t_{\frac{1}{2}}}$

$\dfrac{N}{N_0} = e^{-\frac{\ln 2}{5.27}(14.82)}$

$\quad = 0.14$

Therefore, there is 14% left.

- 2 marks: determines the percentage
- 1 mark: correctly substitutes into a relevant equation

Question 12

a $^{222}_{86}\text{Rn} \rightarrow\ ^{218}_{84}\text{Po} +\ ^{4}_{2}\text{He}$

- 2 marks: provides a correct nuclear equation
- 1 mark: provides a partially correct nuclear equation

b   The mass defect results in the release of energy according to $E = mc^2$, and accounts for the kinetic energy of the products.

- 1 mark: identifies that the mass conversion results in the kinetic energy of the products

Question 13

The tracks for both the positron and electron curve in a circle, which means they must be charged for the magnetic field to exert a force on them.

The forces are in the opposite directions, meaning that they have opposite charges. Because the electron is negatively charged, the positron must be positively charged.

Both have decreasing radii of turn. Due to their decreasing velocity (as they interact with gas molecules and thus lose energy), their radius decreases proportionally, because radius is proportional to velocity ($\frac{mv^2}{r} = qvB$ and therefore $\frac{mv}{r} = qB$).

- 3 marks: explains a property for two tracks
- 2 marks: explains a property for one track
- 1 mark: states any observation in relation to the two tracks seen in the diagram

Question 14

Fusion and fission reactions can both release energy. For both cases, the loss of mass is because of its conversion to energy according to $E = mc^2$.

This energy comes from the mass defect. The combined mass of the products must be less than the combined mass of the reactants. As the graph shows, fusion only occurs in smaller nuclides because fusing them results in a larger binding energy per nucleon, which is as a result of a loss in mass. For fission, an increase in binding energy per nucleon, and thus a loss of mass, only occurs for large nuclides that split.

- 4 marks: explains that for both reactions to occur there needs to be an increased binding energy per nucleon because of a decrease in mass
- 3 marks: outlines that in both reactions there is an increased binding energy per nucleon because of a decrease in mass
- 2 marks: identifies that in both reactions there is an increased binding energy per nucleon or decrease in mass
- 1 mark: provides relevant information

Question 15

The statement is incorrect because it is not possible for a reactor to undergo uncontrolled fission as in a nuclear bomb.

Both a nuclear bomb and a nuclear reactor work on the principle of nuclear fission. A chain reaction occurs when neutrons are released from each individual reaction, which then go on to initiate another nuclear reaction.

However, there are differences between the types of fission reactions in each. A nuclear bomb uses uncontrolled fission of nuclear fuel, in which the rate of reactions is exponential. In a fission reactor, the reaction is controlled using control rods and moderators, and the rate of reaction is more or less linear. The control rods absorb neutrons and thus can prevent further reactions. Moderators slow the neutrons, which may be able to control the rate of reaction.

A second difference is the purity of the fuel. The nuclear fuel in a reactor (such as ^{235}U) is much lower in concentration than in the fuel used in nuclear weapons. The percentage of ^{235}U is low in a nuclear reactor but high in a nuclear bomb.

Even if a reactor has a failure, this would not cause a nuclear explosion. If the moderator drained away, the reaction would stop because the neutrons would no longer be slowed, so the chances of neutron interaction with fuel rods would become greatly diminished.

If control rods froze in the 'out' position, the reactor may overheat. The reaction is still unlikely to be uncontrollable because of the low concentration of fissionable material in the fuel. However, it is possible that a steam pressure rupture might cause a pressure explosion. This would be serious, but far less so than a nuclear explosion.

- 5 marks: provides reasons why a reactor would not explode like a nuclear weapon and makes an assessment of them
- 4 marks: outlines reasons why a reactor would not explode like a nuclear weapon
- 3 marks: identifies differences between a nuclear weapon and a nuclear reactor **and** outlines a reason why a reactor will not explode like a nuclear weapon
- 2 marks: identifies a difference between a nuclear weapon and a nuclear reactor
- 1 mark: provides some relevant information

Question 16   ©NESA   2019 MARKING GUIDELINES SII Q36

Mass defect $= 197.999 - (193.988 + 4.00260) = 0.00084\,u$

Converting to kilograms $= 0.0084 \times 1.661 \times 10^{-27} = 1.395 \times 10^{-29}\,kg$

Total energy produced $= mc^2 = 1.395 \times 10^{-29} \times (3 \times 10^8)^2 = 1.256 \times 10^{-12}\,J$

$KE_{alpha} =$ total energy produced $- KE_{Po} = 1.256 \times 10^{-12} - 2.55 \times 10^{-14} = 1.23 \times 10^{-12}\,J$

Since the radon atom is initially at rest, the decay products move away from each other with equal and opposite momenta. As the alpha particle's mass is significantly less than that of the polonium atom, it therefore has a significantly higher velocity ($p = mv$) and consequently a higher KE. Despite the higher mass of the polonium atom, the higher velocity of the alpha particle has a more significant effect on its KE ($KE = \frac{1}{2}mv^2$).

Answers could include a calculation not using mass defect ($K \rightarrow P$).

- 7 marks: applies the correct method to calculate the kinetic energy of the alpha particle **and** explains the greater kinetic energy of the alpha particle using the principle of conservation of momentum
- 6 marks: applies the correct method to calculate the kinetic energy of the alpha particle **and** applies the principle of conservation of momentum
- 4–5 marks: shows the main steps of the calculation of kinetic energy **and/or** shows a sound understanding of the conservation of momentum
- 2–3 marks: shows step(s) of the calculation of kinetic energy **and/or** shows some understanding of the conservation of momentum
- 1 mark: provides some relevant information

Test 16: Deep inside the atom

Multiple-choice solutions

Question 1

D Proton

A fundamental particle is one that cannot be divided into smaller components. The proton is made up of three quarks and therefore is not a fundamental particle. All of the other particles listed (electron, gluon and muon) are fundamental.

Question 2

B Photons

Particle accelerators manipulate charges using electric and magnetic fields. To do this, the particles need to have an electric charge. Because photons are chargeless particles, they cannot be manipulated with electric or magnetic fields. The other particles listed (electrons, protons and lead nuclei) have a charge.

Question 3

A Electron

A particle with a quark composition *udd* has a charge of zero and is therefore a neutron. A particle with a quark composition *uud* has a charge of +1 and therefore is a proton. In this reaction, a neutron transmutes to a proton. To ensure conservation of charge, there needs to be a release of a negatively charged particle, which is the electron.

C is incorrect because this would only occur for the transmutation of a proton to a neutron. **B** is incorrect because the original particle, *udd*, is a neutron. Producing another neutron in this reaction would violate the conservation of mass. **D** is incorrect because, like **B**, this violates the conservation of mass.

Question 4

B Superconducting magnets produce fields that change the direction of the particles, whereas variations in electric fields increase their energy.

Particle accelerators manipulate charged particles by applying magnetic and electric fields. Magnetic fields are required to change the direction of the charged particles, as the force they experience is perpendicular to their direction of motion. Although this is a form of acceleration, this only changes their direction and not their speed.

Electric fields cause charged particles to accelerate. Because the velocity of the particles increases, so too does their momentum (according to $p = mv$) and energy (according to $KE = \frac{1}{2}mv^2$).

(This is true in low-velocity cyclotrons, but accelerators such SLAC and the LHC accelerate particles to close to the speed of light, so their momentum increases relativistically.)

Question 5

A Lepton: muon; hadron: neutron; boson: gluon

Leptons are fundamental (indivisible) particles. Muons, electrons, neutrinos and positrons are all leptons.

Hadrons are particles made up of a combination of quarks and/or anti-quarks, and thus they are not fundamental particles. Protons and neutrons are examples of hadrons and are made up of three quarks, but hadrons also include a group of particles called mesons, which are made up of a quark and an antiquark.

Bosons are fundamental particles and are force carriers. They include the photon, the gluon and the Higgs boson.

Short-answer solutions

Question 6

Fundamental particles are those that are indivisible. Protons and neutrons are made up of quarks and thus are divisible.

- 2 marks: identifies that protons and neutrons are made up of quarks, and thus are not fundamental particles
- 1 mark: defines a fundamental particle **or** identifies that protons and neutrons are made up of quarks

Question 7

a Quark – fundamental particle, charge of $-\frac{1}{3}$ or $+\frac{2}{3}$

Muon – more massive than the electron, charge of -1, half spin

Positron – same mass as the electron, charge of $+1$

Gluon, photon (bosons) – massless, no charge, spin of 1

Higgs boson – has mass, no charge, zero spin

- 2 marks: identifies a fundamental particle and names one property
- 1 mark: identifies a fundamental particle

b A sample answer is as follows:

A muon is a subatomic particle that has a short half-life and negative charge. It is detected in the study of cosmic rays. These rays strike the upper atmosphere and the subsequent collisions with air molecules produces a cascade of muons, which are then detected in laboratories on the ground.

Other answers could include the evidence for the Higgs boson found using the Large Hadron Collider, and the evidence for the photon seen in the photoelectric effect.

- 3 marks: describes experimental evidence for a subatomic particle
- 2 marks: outlines experimental evidence for a subatomic particle
- 1 mark: identifies experimental evidence

Question 8   ©NESA   2018 MARKING GUIDELINES SII Q34e

In the standard model, fundamental particles – quarks and leptons – interact through four fundamental forces. The constituents of protons and neutrons are three quarks. The strong force holds nucleons together in the nucleus, overcoming the electrostatic repulsion between protons. An electron is a lepton which has no constituents. The force of gravity attracts protons and neutrons within the nucleus but this force is negligible.

The electrostatic force results in a mutual repulsion of protons and an attraction for the electrons toward the nucleus. Electrons orbit the nucleus. The strong nuclear force counteracts the electrostatic force, but only within a limited range of subatomic distances. The weak force is involved in the process of radioactive decay.

- 7 marks: analyses the roles of both forces and particles using the standard model **and** relates these to the current understanding of the atom
- 6 marks: analyses the roles of forces **and/or** particles using the standard model **and** relates these to the current understanding of the atom
- 4–5 marks: describes the roles of forces **and/or** particles in relation to the standard model **and** links these to the current understanding of the atom
- 2–3 marks: shows some understanding of the standard model
- 1 mark: provides some relevant information

Question 9

a $^{18}_{8}O + ^{1}_{1}p \rightarrow ^{18}_{9}F + ^{1}_{0}n + \gamma$

- 2 marks: provides a correct equation
- 1 mark: provides a partially correct equation

b   The protons need high speed to give them the required energy to overcome the repulsive forces of the nuclear target.

- 1 mark: identifies why protons need high speed

c   Protons are injected into the centre of the cyclotron, between the hollow electrode chambers called the dees. The dees are connected to an alternating voltage, so there is an oscillating electric field between them. The protons are attracted first to the negative dee, but this soon becomes positive, repelling the protons, which are already attracted to the opposite dee (which is now negative). This cycle continues at the frequency of the alternating voltage, with the protons accelerating back and forth between the dees under the influence of the oscillating electric field.

The cyclotron is also subjected to a magnetic field. As the protons accelerate back and forth, they turn in a circular path because of the applied magnetic field. As they accelerate, the radius of their circular motion increases, because their velocity is increasing. The net result is that they spiral outward, increasing in speed.

Eventually, when their radius matches the physical radius of the cyclotron, the protons exit at high speed in a straight line towards their target.

- 3 marks: demonstrates a thorough understanding of how the cyclotron accelerates a proton
- 2 marks: shows an understanding of the purpose of the electric field and the magnetic field in a cyclotron
- 1 mark: provides relevant information

Practice HSC exam 1

Multiple-choice solutions

Question 1

B Increasing the intensity of the light source

Increasing the intensity of the light source increases the number of photons. This will increase the number of photoelectrons emitted, thereby increasing the current.

A is incorrect because decreasing the intensity will reduce the current. **C** is incorrect because increasing the frequency will increase the energy and thus the stopping voltage but will not affect the current. **D** is incorrect because decreasing the frequency will decrease the energy and thus the stopping voltage but will not affect the current.

Question 2

D J. J. Thomson

Thomson, by applying electric and magnetic fields to the cathode rays, not only determined their velocity, but was able to determine the charge-to-mass ratio, leading to the discovery of the electron.

A is incorrect because Hertz established evidence for radio waves. **B** is incorrect because Rutherford established the planetary model of the atom. **C** is incorrect because Planck was the first to suggest that energy can be quantised.

Question 3

D

Using the right-hand palm rule, fingers point in the direction of the magnetic field, the thumb points in the direction of the current and the palm gives the direction of the force.

Question 4

D

For this to be projectile motion, the horizontal velocity must be constant and thus the dots must be spaced evenly from left to right. Therefore, **B** and **C** are incorrect. The vertical velocity is not constant as the projectile is accelerating vertically. Therefore, the vertical spacing must increase between the dots; this eliminates **A**.

Question 5

B He provided experimental evidence that electromagnetic waves have a finite speed.

All of the other statements about Maxwell's work are correct statements. Hertz was first to provide experimental evidence for the existence of electromagnetic waves other than light that travelled at the same speed as light.

Question 6

D a shorter range but will curve upwards.

Because the electron is opposite in charge to the proton, it will experience a force in the opposite direction. This makes **A** and **C** incorrect. Although the force the electron experiences is identical to that of the proton (using $F = Eq$), its acceleration will be considerably greater because the electron's mass is significantly less. Therefore, it will have a shorter range.

Question 7

B It is double the size.

$$\frac{r_E^3}{T^2} = \frac{GM}{4\pi^2}$$

$$r_E^3 = \frac{GMT^2}{4\pi^2}$$

$$r_P^3 = \frac{G2M(2T)^2}{4\pi^2} = 8\frac{GMT^2}{4\pi^2} = 8r_E^3$$

$$r_P = 2r_E$$

A is incorrect because the cubed root has not been found. **C** is incorrect because if the satellite is in orbit around a different planet, the radius of the orbit to remain geostationary will not be the same. **D** is incorrect because there is not an inverse relationship between radius and length of day.

Question 8

B 1.23 eV

$$K = hf - \phi = \frac{hc}{\lambda} - \phi$$

The work function is in eV, so we need to calculate the photon energy in eV.

$$E = \frac{\frac{hc}{\lambda}}{1.6 \times 10^{-19}}$$

$$= \frac{\left(\frac{(6.626 \times 10^{-34})c}{200 \times 10^{-9}}\right)}{1.6 \times 10^{-19}}$$

$$= 6.21\,\text{eV}$$

$$K = 6.21 - 4.98$$

$$= 1.23\,\text{eV}$$

A is incorrect because only the energy of the photon has been calculated. **C** is incorrect because this is the work function. **D** is incorrect. Because the incoming photon has a greater energy than the work function, the photoelectron will have kinetic energy.

Question 9

C 2 : 1

The velocity of the pendulum can be determined in relation to the radius and frequency.

$$v = \frac{2\pi r}{T} = 2\pi rf$$

$$v^2 = 4\pi^2 r^2 f^2$$

However, as it is a conical pendulum, $v^2 = gr\tan\theta$

Combining the two:

$$gr\tan\theta = 4\pi^2 r^2 f^2$$

$$\tan\theta_r = \frac{4\pi^2 r^2 f^2}{g}$$

$$\tan\theta_{2r} = \frac{4\pi^2 2r^2 (0.5f)^2}{g} = \frac{4\pi^2 r^2 (f)^2}{2g} = \frac{\tan\theta_r}{2}$$

A is incorrect as a result of not squaring the frequency change. **B** is incorrect because the ratio is in the wrong order. **D** is incorrect because an incorrect substitution has been made.

Question 10

B 2.57×10^{13} kg

The energy that the Sun releases is from the conversion of mass during fusion.

$$E = Pt = mc^2$$
$$3.85 \times 10^{28} \times 60 = m(3 \times 10^8)^2$$
$$m = 2.56 \times 10^{13} \text{ kg}$$

A is incorrect and is the amount of mass lost each second. **C** is incorrect because it is the amount of mass lost each second, as well as the calculation failing to square c. **D** is incorrect and is a result of failing to square c.

Question 11

D 1.85 m s^{-1}

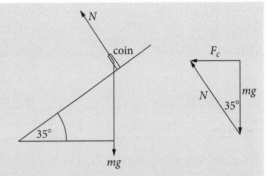

Two forces are acting on the coin: the normal (N) and the force due to gravity (mg). The **net** force is the centripetal force (F_c).

$$\tan\theta = \frac{F_c}{mg} = \frac{mv^2}{rmg} = \frac{v^2}{rg}$$
$$v^2 = gr\tan\theta$$
$$= (9.8)(0.5)\tan 35$$
$$v = 1.85 \text{m s}^{-1}$$

A is incorrect because the square root was not taken. **B** is incorrect because the radius of 50 cm was not converted to metres. **C** is incorrect because the radius of 50 cm was not converted to metres and the square root was not taken.

Question 12

A

V_1 is the primary coil and V_2 is the secondary coil. When the switch is closed for the first 2 s, the graph should show a steady voltage. When the switch is opened, there is a momentary change in flux as the magnetic field in the primary coil drops to zero. This causes a brief spike in voltage in the secondary coil as it experiences this changing flux according to Faraday's law. Because it is a negative change in flux, according to Lenz's law this will result in a positive voltage in the secondary coil.

D is incorrect because it has the voltage switched the other way round. **B** and **C** are incorrect because they show a voltage produced in the secondary coil when there is no change in flux in the primary coil.

Question 13

A

Because 4.3 light-years is measured from Alex's perspective, the proper time is:

$$t = \frac{d}{v} = \frac{4.3 \times 2}{0.6c} = 14 \text{ years, so Alex will be 39 upon Toni's return.}$$

Toni's time will be less because of time dilation. Toni's time is also less because of the length contraction from her perspective.

B is incorrect because it only accounts for half the journey. **C** and **D** have the ages the wrong way round, and **D** only accounts for half the journey.

Question 14

B $2.9 \times 10^8 \, \mathrm{m\,s^{-1}}$

The diagram shows 1.5 wavelengths for the given 8.7 m. This means that the wavelength is $\dfrac{8.7}{1.5} = 5.8 \, \mathrm{m}$

$v = f\lambda = (50 \times 10^6)(5.8) = 2.9 \times 10^8 \, \mathrm{m\,s^{-1}}$

A is incorrect because the calculations do not give this value; instead, this is the accepted value of 'c', the speed of light. **C** and **D** would result if you assumed that the speed of light is c and calculated the wavelength mistakenly.

Question 15

C When electrons move between energy levels, they release energy with a specific wavelength. The arrows represent that movement, and the larger the arrow, the smaller the wavelength.

According to Bohr, an electron can only exist in discrete energy levels. This means that if the electron moves down energy levels, it will release energy of a specific wavelength and corresponding frequency.

The wavelength can be determined by the Rydberg formula:

$$\frac{1}{\lambda} = R\left(\frac{1}{n_f^2} \frac{1}{n_i^2}\right)$$

The energy will be based on $E = hf = \dfrac{hc}{\lambda}$.

Thus the larger the jump in energy levels (i.e. the greater the difference between n_f and n_i), the smaller the wavelength of the photon emitted and therefore the greater amount of energy.

A is incorrect because the larger arrows mean more energy and therefore smaller wavelengths.

B is incorrect because if energy of a specific value is received, the electron energy level will increase, not decrease.

D is incorrect because the amount of energy depends not on the initial energy level but on the difference between energy levels, according to Rydberg's formula.

Question 16

D eddy currents will cause the metallic materials to move more slowly than other materials.

As non-magnetic metals such as aluminium and copper move over the magnets, eddy currents are produced. These produce magnetic fields that oppose the relative motion, consistent with Lenz's law. This causes them to move more slowly. This is consistent with the law of conservation of energy.

A is incorrect because aluminium and copper are not magnetic.

B is incorrect because, although eddy currents are produced, they will cause a repulsion, not an attraction.

C is incorrect because the magnetic fields will cause them to move more slowly. If they moved more quickly, this would violate the law of conservation of energy.

Question 17

C Star A: cooler, larger radius; star D: hotter, smaller radius

A and **B** are correct, but **C** is a better description.

D is incorrect because the lifespans of stars A and D are similar. D is just further along in its life.

Question 18

C

For an EMF to be generated, the charged particles within the conductor must move perpendicular to the magnetic field. This does occur (at a minimum when PQ is vertical and a maximum when PQ is horizontal). However, the direction of the EMF will be into or out of the page. Because this is not along the conductor, no real EMF will be generated between P and Q.

Question 19

C 5.6×10^{-9} m

The wavelength of the electron can be determined by de Broglie's relationship, where $\lambda = \dfrac{h}{p}$.
However, because of momentum dilation, $p = \gamma mv$.

$$p = \frac{mv}{\sqrt{1 - \dfrac{v^2}{c^2}}} = \frac{(9.109 \times 10^{-31})(0.4c)}{\sqrt{1 - (0.4^2)}} = 1.19 \times 10^{-22}\,\text{N s}$$

$$\lambda = \frac{h}{p} = \frac{6.626 \times 10^{-34}}{1.19 \times 10^{-22}} = 5.56 \times 10^{-12}\,\text{m}$$

Because $\lambda = d\sin\theta$, where $\sin\theta$ approximates to y/L:

$$d = \frac{\lambda L}{y} = \frac{5.56 \times 10^{-12}(1)}{1 \times 10^{-3}} = 5.56 \times 10^{-9}\,\text{m}$$

A is incorrect because it is quoting the momentum. **B** is incorrect because it fails to account for relativistic behaviour. **D** is incorrect because it is quoting the wavelength.

Question 20

D   Energy is released by the reaction because the binding energy of the products is greater than the sum of the binding energies of the reactants.

To determine whether the reaction will release energy or not, compare the mass of the reactants with the mass of the products or examine the binding energy per nucleon of the species.

In this case, regardless of what Z is, there is a significant increase in the binding energy (and binding energy per nucleon) of the products.

W + X (= 2.224 566 + 8.481 798) < Y + Z (= 28.295 66 +??)

This means there was an increase in the mass defect, so energy must be released according to $E = mc^2$.

Therefore, the answer is **D**.

A would be correct if the question said, '... the mass defect of Y is greater than the *sum* of both W and X', because any mass defect of Z could only add to the mass defect on the right of the equation. **B** is incorrect because it *can* be deduced. **C** is incorrect because it states that energy is input rather than released.

Short-answer solutions

Question 21

a $E = hf$
$$= (6.62 \times 10^{-34})(8.0 \times 10^{14})$$
$$= 5.3 \times 10^{-19} \text{ J}$$

- 2 marks: correctly calculates the energy of a photon of UV light
- 1 mark: substitutes into a relevant formula

b   The student shines lights of different frequencies on the zinc plate. The electroscope is charged, as can be seen by the gold leaves repelling each other. When the red light is shone on the plate, irrespective of the charge on the plate, there is no change in the position of the leaves. This suggests that the energy from the red light is below the threshold frequency.

When the UV light is shone on the positively charged electroscope, no change is seen because more electrons are being emitted, making the plate more positive. Therefore, the leaves stay apart.

When the UV is shone on the negatively charged plate, the leaves move closer together. This suggests that the energy is above the threshold frequency, giving the electrons enough energy to be ejected from the surface of the zinc. As the electrons are ejected, the charge on the electroscope decreases, which results in the leaves coming together.

- 3 marks: explains the conclusion based on observations from the diagram
- 2 marks: relates observations to theory
- 1 mark: makes observations

Question 22

a $m_2 > m_1$

- 1 mark: states $m_2 > m_1$

b $qvB = \dfrac{mv^2}{r}$

$$r = \dfrac{mv}{qB}$$

$$\Delta r = r_2 - r_1 = \dfrac{m_2 v}{qB} - \dfrac{m_1 v}{qb}$$

$$= \dfrac{v}{qb}(m_2 - m_1)$$

- 3 marks: correctly derives Δr
- 2 marks: attempts to determine Δr
- 1 mark: identifies the correct formula

Question 23

Particle accelerators use the property of charge to manipulate charges to very high speeds.

A good example is the Large Hadron Collider (LHC). This uses electric fields and magnetic fields to manipulate the motion of charged particles.

In the case of the LHC, protons (hydrogen atoms that have been ionised) are accelerated by passing them through a series of chambers that have strong electric fields. They experience a force according to $F = Eq$. Each chamber increases the speed of the protons.

The protons are then made to travel in a circular path by passing them through a large ring containing

strong magnetic fields. These magnetic fields ensure the protons remain in a circular path, according to

$F = qvB = \dfrac{mv^2}{r}$. Eventually, the protons return to the electric-field chambers, where the process is repeated.

Each turn of the loop will result in an increase in the velocity, and therefore the energy, of the protons.

Because the radius of the path remains constant, the magnetic field must be adjusted for each increase in velocity (according to $r = \dfrac{mv}{qB}$). All of this links to the concepts of electromagnetism.

As the velocity of the protons increases to relativistic speeds, their momentum will dilate according to:

$$p_v = \frac{p_0}{\sqrt{1 - \dfrac{v^2}{c^2}}}$$

This will also increase the energy of the protons. This is consistent with concepts of relativity.

The protons are then collided with other protons travelling the in opposite direction. They annihilate, and the resulting energy converts to the mass of the new subatomic particles according to $E = mc^2$.

Other examples include linear accelerators such as SLAC and LINACs.

> - 5 marks: using a named example, demonstrates a thorough understanding of the physics of electromagnetism and relativity in particle accelerators
> - 4 marks: demonstrates a thorough understanding of the physics of electromagnetism and relativity in particle accelerators
>   **or**
>   using a named example, demonstrates a sound understanding of the physics of electromagnetism and relativity in particle accelerators
> - 3 marks: outlines some physics concepts of electromagnetism and relativity
>   **or**
>   demonstrates a sound understanding of the physics of electromagnetism **or** relativity in particle accelerators
> - 2 marks: identifies a physics concept of electromagnetism **or** relativity in the working of a chosen particle accelerator
> - 1 mark: provides relevant information

Question 24

The rope represents an electromagnetic wave such as visible light. The fence represents the polariser.

The model correctly identifies that only a specific orientation of the wave would be transmitted: in this case, where the wave is parallel to the gaps in the fence (which represent the polariser plane).

However, there are several weaknesses with this model.

Whereas the rope has only one plane of vibration, an electromagnetic wave has many, along different axes.

The rope's wave will not pass through effectively if it is not completely parallel to the polariser, because of friction between the rope and the fence. This is not consistent with Malus' law. In contrast, if there is a component of an electromagnetic wave that is parallel to a polariser, some of the wave will be transmitted but its intensity will be reduced, consistent with Malus' law.

Both waves are transverse waves. However, another weakness with the model is that the wave represented requires a medium (the rope) to travel, whereas an electromagnetic wave requires no medium and is a result of fluctuations of electric and magnetic fields.

> - 4 marks: describes the strength **and** weaknesses of the model
> - 3 marks: outlines a strength **and** a weakness of the model
> - 2 marks: identifies a strength **and** a weakness of the model, **or** outlines a strength **or** a weakness of the model
> - 1 mark: provides some relevant information

Question 25

a   Allow for small variations in how the graph is drawn.

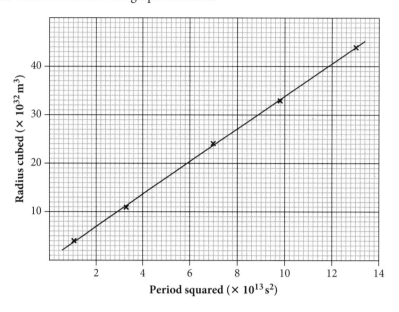

> • 2 marks: correct plotting with line of best fit
> • 1 mark: correct plotting

b   Kepler's third law is a linear relationship between the variables r^3 and T^2. Because there is a strong linear relationship in the graph, the results are valid.

> • 3 marks: assesses the validity of the graph with reference to Kepler's third law
> • 2 marks: identifies that $\dfrac{r^3}{T^2}$ is linear, relating it to Kepler's third law
> • 1 mark: identifies Kepler's third law or identifies that $\dfrac{r^3}{T^2}$ is linear

c $\text{slope} = \dfrac{r^3}{T^2} = \dfrac{GM}{4\pi^2}$

Using $(10, 33)$ and $(4.5, 15)$

$$\frac{(33 - 25) \times 10^{32}}{(10 - 4.5) \times 10^{13}} = \frac{18 \times 10^{32}}{5.5 \times 10^{13}} = \frac{(6.67 \times 10^{-11})m}{4\pi^2}$$

$$m = \frac{18 \times 10^{32}}{5.5 \times 10^{13}} \frac{4\pi^2}{(6.67 \times 10^{-11})} = 1.9 \times 10^{31} \, \text{kg}$$

> • 2 marks: determines the mass from the slope
> • 1 mark: determines the mass from a single value or attempts to use the slope to determine the mass

Question 26

a   The motorcycle must experience a resultant force inward to cause circular motion. This is the centripetal force, determined by $F = \dfrac{mv^2}{r}$, and is a result of the normal force acting inward to the centre of the sphere and the weight of the motorcycle.

There must also be an upward vertical component of the normal to cancel out the force due to gravity, for the riders to 'defy gravity'. This means, ignoring any frictional forces, it is not possible to perform this manouvre **at** the equator, where there would be no vertical component to the normal force to cancel out the gravitational force. The magnitude of the centripetal force is dependent on velocity, and so the motorcyclist must reach a minimum speed to succeed.

- 3 marks: explains balanced forces and the need for a particular speed to be met
- 2 marks: describes the forces involved
- 1 mark: identifies some relevant information

b $F_N = \dfrac{mv^2}{r} + mg$

$0 = \dfrac{v^2}{r} + g$

$\dfrac{(v)^2}{4.86} = 9.8$

$v = 6.9\,\mathrm{m\,s}^{-1}$

- 2 marks: correctly calculates the value of v
- 1 mark: substitutes into a relevant formula

Question 27

As a motor starts to turn because of the motor effect, the coil rotates in a magnetic field. When the switch is closed, the supplied voltage in the circuit, equal to the net voltage across the motor, is at a maximum. This can be seen on the graph when $t = 0\,\mathrm{s}$.

As the motor rotates, the coils cut through magnetic flux and act as a generator. Because of the rate of change of flux, an EMF is induced which opposes the rotation – the back EMF. As this is induced in the opposite direction to the supplied voltage, the net voltage available to the motor is decreased. The magnitude of the back EMF increases as the speed of the motor increases, so the net voltage is reduced as speed is increased (net voltage = supply voltage – back EMF), as evidenced by the trends of the two graphs.

Because the net voltage is decreasing, the current decreases and therefore the rate of change of speed (acceleration) of the motor decreases.

Some work is always done against friction and any other work must be added to this, meaning that the net voltage will reduce to a constant non-zero value and will produce a constant speed of rotation until more or less work is required of the motor. Hence, the voltage at $0.05\,\mathrm{s}$ is not zero.

(If there was no load on the motor, which includes no friction, the net voltage would reach zero. This does not occur.)

- 4 marks: explains the graph with the presence of back EMF
- 3 marks: describes the effects of back EMF
- 2 marks: identifies the presence of back EMF
- 1 mark: provides relevant information

Question 28 ©NESA 2020 MARKING GUIDELINES SII Q27 (ADAPTED)

A sample answer is as follows.

For diffraction, $d \sin \theta = m\lambda$, where $m = 1$ for maxima A, B.

Hence $\sin \theta = \dfrac{\lambda}{d}$ and so decreasing λ decreases θ.

To keep the distance between A and B constant:

- the slit separation d could be reduced, causing θ to increase, compensating for the effect of reducing the wavelength

- reducing the wavelength will cause $\sin \theta$ to decrease, and thus the distance between A and B decreases. Increasing the distance (y) would counteract this change because $\sin \theta$ approximates to $\tan \theta$ ($= \dfrac{AB}{y}$).

> - 4 marks: explains two methods
> - 3 marks: explains one method and identifies another **or** outlines two methods
> - 2 marks: outlines one method **or** identifies two methods
> - 1 mark: provides some relevant information

Question 29

Nuclear fission is the process of splitting a large nucleus into smaller parts and releasing energy. As seen in the equation, this process involves the capture of a neutron by the uranium nucleus.

If the neutron moves at the correct speed, it will be captured. The fission products will be a barium nucleus, a krypton nucleus and three more neutrons. These neutron products are important when determining whether the reaction is controlled or uncontrolled.

Whether it is part of a controlled or uncontrolled reaction, each fission of a uranium nucleus follows the same reaction and so releases the same amount of energy.

Mass of reactants $= 235.0439\,u + 1.008\,665\,u = 236.052\,565\,u$

Mass of products $= 140.9144\,u + 91.9263\,u + 3 \times 1.008\,665\,u = 235.866\,695\,u$

This is a loss of $0.18587\,u$.

Therefore, the energy released is $0.18587\,u \times 931.5\,\text{MeV}\,c^{-2} = 173.14\,\text{MeV}$ per reaction.

In an uncontrolled reaction, more than one of the three daughter neutrons will go on to hit and split another uranium nucleus. If there is enough uranium, this will continue in a chain reaction. The nature of a chain reaction means that the number of reactions per second increases exponentially, and so the energy available from the mass of uranium is released almost instantly, resulting in an explosion.

For an uncontrolled reaction to occur, the density of uranium nuclei must be sufficient such that more than one of the three neutrons released will be absorbed by another uranium nucleus. The minimum amount of uranium for this condition is known as the critical mass. The lowest critical mass occurs when the uranium is a spherical shape.

For a controlled reaction, the number of neutrons available to initiate further fission must be controlled. If only one of the three neutrons produced is, on average, available to initiate further fission, then energy from the uranium mass will be released at a constant or controlled rate. The same amount of energy is released from the same amount of uranium in both reactions, but in a controlled reaction it is released over a vastly longer time than in an uncontrolled reaction.

In a nuclear reactor, for example, some of the neutrons that are produced by each reaction are absorbed by the control rods and therefore are not available to hit the fuel rods (where the uranium is). This makes fewer neutrons available for further reactions, thereby controlling the number of reactions and the amount of energy released per second.

- 5 marks: extensively compares controlled and uncontrolled fission reactions in terms of the energy output and the processes involved
- 4 marks: thoroughly compares controlled and uncontrolled fission reactions in terms of the energy output and the processes involved
- 3 marks: compares controlled and uncontrolled reactions in terms of the processes involved
- 2 marks: compares the energy or processes involved
- 1 mark: provides some relevant information

Question 30

a $\tau = nIAB$

$= (3)(1.5)(0.10 \times 0.05)(0.3)$

$= 6.75 \times 10^{-3}\,\text{N m}$

- 2 marks: correctly determines the torque
- 1 mark: makes a partially correct substitution into a correct equation

b   To convert the DC motor to an AC motor, the split-ring commutator needs to be replaced by slip rings. This will change the direction of the current within the rotating commutator as the AC supply changes direction.

However, this does introduce a limitation: the motor's rotational speed is fixed to the frequency of the supply voltage, hence it will only rotate at a constant rate, unlike the DC motor, which can have a variable rotational speed. (This assumes there is no load on the motor.)

- 3 marks: explains a modification to make it work as an AC motor **and** assesses its effectiveness
- 2 marks: explains a modification to make it work as an AC motor
- 1 mark: identifies a modification to make it work as an AC motor

Question 31

The two experiments are similar in that they both represent forces caused by fields, where the resulting circular motions for both are the result of repulsive forces that cause a rotation in the same direction as the force applied.

However, this is where the similarity ends.

In experiment A, the two charges are identical in sign and therefore both have an electric field that radiates outwards. As the smaller ball approaches the larger ball, it experiences a force of repulsion, according to Coulomb's law, where the force is inversely proportional to the square of the distance between them.

The large ball also experiences a force (consistent with Newton's third law) and accelerates away. This makes the disc turn in the same direction as the applied force.

Experiment B is a result of magnetic fields rather than electric fields.

As the magnet moves across the disc it creates eddy currents in the disc, both in front of and behind the magnet. The eddy current in front rotates in an anticlockwise direction (seen from above), consistent with Lenz's law. This produces a north pole above the disc (using the right-hand grip rule). The result is a force of repulsion between the north pole of the magnet and the north pole of the eddy current.

Similarly, an eddy current is produced behind the magnet – in this case, a clockwise eddy current with a south pole above the disc. The result is a force of attraction between the two.

The resulting repulsion in front and attraction behind causes the disc to turn in the same direction as the motion of the magnet.

- 7 marks: provides a thorough understanding of the physics principles of both experiments, comparing the two experiments
- 6 marks: provides a thorough understanding of the physics principles of both experiments
- 5 marks: provides a sound understanding of the physics principles of both experiments with an attempt at comparisons
- 4 marks: provides a sound understanding of the physics principles of both experiments **or** identifies a physics principle for experiments A and B and attempts a comparison
- 3 marks: identifies a physics principle for experiments A and B **or** outlines a physics principle for an experiment
- 2 marks: identifies a physics principle for experiment A or B
- 1 mark: provides relevant information

Question 32 ©NESA 2021 MARKING GUIDELINES SII Q32

As the elastic is stretched, each marking moves by an amount proportional to its distance from the fixed end. This successfully models Hubble's observation that more-distant galaxies are receding from Earth at greater rates, proportional to their distance from Earth.

However, a limitation of the investigation is that it can only model expansion in one dimension, whereas Hubble's evidence showed that the Universe is expanding in three dimensions.

Answers could include:

Other benefits or limitations of the model, such as:

- The investigation only examines a change in position of the markings, while Hubble measured the speed of galaxies.

- As the elastic stretches, the markings themselves will stretch. Hubble did not find that the galaxies were spreading out.

- 5 marks: justifies how well the observations model Hubble's evidence
- 4 marks: provides some justification for how well the observations model Hubble's evidence
- 2–3 marks: relates aspects of the investigation to the expansion of the Universe
- 1 mark: provides some relevant information

Question 33

Bohr devised a planetary model of the atom, in which the electrons existed in discrete energy orbits. He hypothesised that when electrons moved between discrete energy levels, they would release a discrete amount of energy ($E = hf$). He was able to offer a reason why Rydberg's formula could predict the wavelength of light emitted in emission spectra, by arguing that the n values were the energy levels.

Although this model was able to describe spectral phenomena, it was not able to explain why electrons have discrete energy values. He also proposed the quantisation of angular momentum, where the angular momentum $L = n\left(\dfrac{h}{2\pi}\right)$.

Louis de Broglie argued that electrons could have discrete orbits if they were treated as standing waves surrounding the nucleus. Because standing waves have discrete frequency values $\left(f = \dfrac{nv}{2L}\right)$, and thus discrete energies ($E = hf$), he was able to explain why electrons were in stable orbits. This resulted in de Broglie combining the momentum of the electron with its wavelength: $\lambda = \dfrac{h}{p}$.

Question 34

When the motor is switched on, a current is applied to the motor and, because of the motor effect, it will start to turn ($F = BIl$). This creates a torque, τ, where $\tau = nIAB = Fd_1$, that lifts the arm.

For a mass to be lifted, the torque applied by the motor must be equal to or greater than the torque applied by the weight of the mass ($\tau = mgd_2$).

There are several factors that limit the ability of the machine to lift a mass:

- The placement of cable attachment – if d_1 were decreased, this would decrease the torque generated, which would limit the amount of mass that could be lifted. Increasing d_1 could increase the torque, but it would also result in a decrease in angle, thereby limiting the device according to $\tau = mgd_1 \sin \theta$. Also, moving the mass further from the pivot (increasing d_2) would increase the torque required.

- As the arm is lifted, the angle between the cable and the arm decreases, and because $\tau = mgd_1 \sin \theta$, the applied torque decreases. However, so does the torque due to the mass, because $\tau = mgd_2 \sin \theta$. The end result is that the decreasing angle alone is not a limiting factor.

- The DC motor coils have a radius and if this were to be increased by replacing it with another motor, this could increase the ability to lift more mass because torque would be increased.

- Increasing the voltage will increase the current, thereby increasing the torque. However, there are two limitations. First, an increasing current would cause heating, which would increase resistance and thus limit the amount of current that could be generated. Second, an increase in the speed of rotation would increase the back EMF, limiting the amount of current that could be generated.

Question 35 ©NESA 2021 MARKING GUIDELINES SII Q34

a   After the mass is launched at t_0, downwards gravitational acceleration reduces the vertical velocity of the mass until it is zero, at time t_1. The KE is a minimum here, but it is not zero since the horizontal component of its velocity is unaffected by gravity. After t_1, the kinetic energy of the mass increases again as the vertical component of motion increases under the influence of gravity until it strikes the ground at t_2. The kinetic energy at t_2 is greater than the value at t_0 because the mass has lower GPE than at t_0.

b At t_0, find u_y:

$$KE = \frac{1}{2}mv^2$$

$$v^2 = 2\frac{(864)}{3} = 576$$

$$v = 24\,\text{m}\,\text{s}^{-1}$$

Using Pythagoras,

$$24^2 = u_y^{\,2} + 13.76^2$$

$$u_y = 19.66\,\text{m}\,\text{s}^{-1}$$

At t_2 find v_y:

$$KE = \frac{1}{2}mv^2$$

$$v^2 = 2\frac{(1393)}{3} = 928.7$$

$$v = 30.47\,\text{m}\,\text{s}^{-1}$$

Using Pythagoras,

$$30.47^2 = v_y^{\,2} + 13.76^2$$

$$\text{So}\quad v_y = 27.19\,\text{m}\,\text{s}^{-1}$$

To calculate time of flight:

$$v_y = u_y + a_g t$$

$$t = \frac{(v_y - u_y)}{a_g}$$

$$= \frac{27.19 - (-19.66)}{9.8}$$

$$= 4.8\,\text{s}$$

Answers could include calculations using $\Delta u = mg\Delta h$.

- 3 marks: correctly calculates time of flight
- 2 marks: shows some working to calculate time of flight
- 1 mark: provides some relevant information

Practice HSC exam 2

Multiple-choice solutions

Question 1

C

In projectile motion, the horizontal velocity is constant so the object's displacement for each time interval is equal. When plotted, this gives a straight line with a non-zero slope.

The vertical velocity is not constant — the displacement changes with the square of the time interval. When plotted, this gives a parabolic shape.

Only the graphs in option **C** fit these descriptions.

Question 2

D Polarity of magnet 1: north; polarity of battery: negative

Because there is repulsion, the polarity of 1 is north. Using the right-hand grip rule, the current must go up the coil around the soft iron core.

This means the polarity of battery terminal 2 is negative.

Question 3

D Rotating the loop by 45° about axis BC

There is no flux in its current position, according to $\phi = BA\cos\theta$, where $\theta = 90°$ (the angle between the surface vector and the magnetic field lines).

Therefore, increasing neither B nor A will change this, so options **A** and **B** are incorrect.

Rotating the loop about AB will also not change θ, and the flux will remain at zero. Only rotating it about BC will increase the number of flux lines passing through the area and thus will generate an EMF.

Question 4

B $1.0c$

According to Einstein, the speed of light is the same for all observers, irrespective of their relative motion. Therefore, any observer will always measure the speed of light as c. **A**, **C** and **D** are incorrect because they state a change in speed.

Question 5

B Volume

Particles in accelerators are manipulated using electric and magnetic fields. This means that they must carry a charge. The amount of acceleration they experience, which either speeds them up or changes their direction, will be affected by their mass and their speed.

Question 6

D Newton believed that light could travel only as a particle, whereas Huygens believed that light could only travel as a wave.

Newton did not consider that light travelled as a wave; hence, **A** and **C** are incorrect. (In fact, the idea that light behaves differently depending on the experiment did not arise until the 20th century.)

Question 7

A Source of spectrum: reflected sunlight; features labelled Y: absorption lines

The general shape is one of a black body curve, meaning that the object emits a range of wavelengths at different intensities. This means that it cannot be a discharge tube because discharge tubes emit specific wavelengths of electromagnetic radiation; this rules out **B** and **D**. Specific wavelengths are missing from the curve, which means the light coming from the source must have passed through a material that has absorbed specific wavelengths, corresponding to absorption lines.

Question 8

D Light is a wave

Although this is correct, it was already known that light was a wave through the work of Young's double-slit experiment.

A and **C** are incorrect responses because Maxwell's equations showed mathematically that light travelled at a constant speed of $3 \times 10^8\,\mathrm{m\,s^{-1}}$. The mathematics suggested that there were other waves that also did this, which led to the discovery of the electromagnetic spectrum. **B** is an incorrect response because although it was known that changes in electric fields give rise to changing magnetic fields, it was Maxwell who understood that the resulting change in magnetic fields resulted in changing electric fields, which led to the idea that these waves were self-propagating.

Question 9

A $5.0 \times 10^3\,\mathrm{m\,s^{-1}}$

$$F_E = F_B$$

$$Eq = \frac{V}{d}q = qvB$$

$$v = \frac{V}{Bd} = \frac{120}{(1.2)(2.0 \times 10^{-2})} = 5.0 \times 10^3\,\mathrm{m\,s^{-1}}$$

B is incorrect because correct SI units have not been used. **C** and **D** are incorrect because an incorrect substitution has been made.

Question 10

D 3 : 2

The ratio of the primary to secondary voltages is the same as the ratio of turns in the primary coil to turns in the secondary coil.

A and **C** are incorrect because the ratio is in the wrong order. **B** shows incorrect working to get the ratio.

Question 11

B ^{4}He is produced from the cycle.

A is incorrect because although ^{12}C is produced in the CNO cycle, it is reused and therefore not a product of the net reaction. **C** is incorrect because energy is released from this reaction. **D** is incorrect because helium is produced from the cycle.

Question 12

C Wavelength: shorter; spectral radiance: higher

The Sun is a G-type star. An O-type star is significantly hotter. According to Wien's law, its peak wavelength will be shorter. The intensity, or spectral radiance, of this wavelength is also significantly higher.

A is incorrect because the irradiance will increase in an O-type star. **B** and **D** are incorrect as the wavelength will decrease in an O-type star.

Question 13

C The EMF would all be positive, but the flux would remain the same.

A and **D** are incorrect because the only difference between an AC generator and a DC generator is the replacement of the slip rings with a split-ring commutator.

This means that the amplitude and period of the EMF would be identical, as would the magnitude of the EMF produced. However, the EMF would all be positive, so the resulting current would be in one direction only, as shown in the graph below.

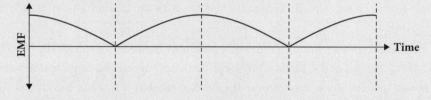

B is incorrect because the flux is independent of what happens to the output and is only dependent on the magnetic field and orientation of the loop. Therefore, it would remain unchanged.

Question 14

C It is turning right.

The ball would normally travel straight if the train were an inertial frame of reference (i.e., if the train were stationary or moving at a constant velocity).

However, the ball travels in a curved path. This means that the train is a non-inertial frame of reference, so it is accelerating. The ball's path appears to be curving because of its own inertia; the ball would still be travelling straight in its frame of reference. However, the ball is viewed from the train's frame of reference and thus if the ball's path curves down (as seen in the diagram), the train's path curves up. This means it is accelerating in the upwards direction. According to the diagram, this is consistent with the train turning right with the centripetal acceleration initially going up the page.

B and **D** are incorrect because both those scenarios would not affect the ball's trajectory, only its relative speed. **A** is incorrect because the ball would curve upward in the diagram if the train were turning left.

Question 15

D None of these would cause an electron to move from $n = 3$ to $n = 5$.

The minimum energy for this transition is 1.55×10^{-19} J:

$$\frac{1}{\lambda} = R\left(\frac{1}{n_f^2} - \frac{1}{n_i^2}\right) = (1.097 \times 10^7)\left(\frac{1}{5^2} - \frac{1}{3^2}\right)$$

$$\lambda = -1.28 \times 10^{-6}\,\text{m}$$

$$E = hf = \frac{hc}{\lambda} = \frac{(6.626 \times 10^{-34})(3.0 \times 10^8)}{-1.28 \times 10^{-6}} = -1.55 \times 10^{-19}\,\text{J}$$

The negative symbol tells us that work is done on the electron and so energy is absorbed.

For an electron to move between energy levels, the energy of the photon must be equal to the difference in those energy levels. Therefore, options **A**–**C** would result in the photon passing through the atom and not causing any transition.

The value for **A** is more than this amount, and the value for **B** is less than this amount. This makes both incorrect because the energy of the photon must match exactly for the transition to occur.

C is a result of failing to square the n values in Rydberg's formula and thus is incorrect. The energy of this photon is more than 1.55×10^{-19} J – this is too much energy for a transition from $n = 3$ to $n = 5$ and would result in the electron being removed from the atom.

Question 16

C B direction at X: right; needle direction: left

When the switch is closed, the current will enter the coil 1. This will produce a magnetic field that will travel through the coil towards the left (using right-hand grip rule).

Coil 2 will experience this field, but according to Lenz's law will produce a field that opposes it. This means the field goes towards the right. Therefore, the field is towards the right at X.

The current that is induced will be such that it leaves coil 2 at its left end. Therefore, it will enter the ammeter at the negative terminal. Because the convention is that the current must enter at the positive terminal, the needle will move to the left.

A and **B** are incorrect most likely because the hand rule has been used incorrectly. **D** is incorrect most likely because the polarity of the ammeter has not been recognised.

Question 17

B t_Q

The period is measured using the observation of when Io moves behind Jupiter and then when it reappears from behind Jupiter. In essence, it is an eclipse.

The time is recorded between those two events, and the difference between them allows the observer to determine the period. However, the observer must also factor in the time taken for the light from Io and Jupiter to reach the observer on Earth.

Therefore, the measured period would increase as Earth moved away from Jupiter and decrease as Earth approached Jupiter. This means that, theoretically, $t_P < t_S < t_Q < t_R$. ($t_S < t_Q$ because the second measurement is closer to Jupiter.)

In reality, t_R cannot be measured because the Sun is blocking the path of the light.

Question 18

B

As the copper is moving to the right, and with the magnetic field going into the copper, the copper to the right of the magnet will produce an eddy current so that its magnetic field will oppose the motion. Because the copper is moving away from the magnet, it will produce a south pole closest to the magnet that will attract it towards the magnet. This will result, on the right side of the magnet, an eddy current in the clockwise direction (according to the right-hand grip rule).

Conversely, on the left side of the magnet, the eddy current will produce a magnetic field that causes a repulsion, because the copper is entering the magnetic field. This means that this eddy current will be anticlockwise.

A and **D** are incorrect because the eddy currents in the reverse directions would result in the copper speeding up. This would violate the law of conservation of energy. **C** is incorrect because both eddy currents are in the same direction, which is not possible.

Question 19

D 82%

First, determine the time taken for a muon to travel the distance given.

$$t = \frac{d}{v} = \frac{100}{(0.6)(3.0 \times 10^8)} = 5.56 \times 10^{-7} \, s$$

To determine the ratio, first find the dilated half-life of the muons.

$$t_v = \frac{t_0}{\sqrt{1 - \frac{v^2}{c^2}}} = \frac{1.56 \times 10^{-6}}{\sqrt{1 - (0.6)^2}} = 1.95 \times 10^{-6} \, s$$

Therefore, the proportion of muons detected is:

$$\frac{N}{N_0} = e^{-\lambda t} = e^{\frac{-\ln 2(t)}{t_{\frac{1}{2}}}} = e^{\frac{-\ln 2(5.56 \times 10^{-7})}{1.95 \times 10^{-6}}} = 0.82$$

As a percentage, this is 82%.

A is incorrect and is a result of dividing the time taken by the dilated half-life. **B** is incorrect and is a result of dividing the time taken by the proper half-life. **C** is incorrect and is a result of correctly using the half-life formula but using the proper time instead of the dilated half-life.

Question 20

B $2hf_1$

Each photon falling on the plate has an energy of $E = hf$.

No photoelectrons are emitted until the frequency of the light reaches f_1. When the frequency of the light reaches f_2, photoelectrons are emitted from both metal 1 and metal 2.

Because energy input = energy output, then:

$hf = \phi + KE$

When the frequency $= f_1$, KE $= 0$ for metal 1,

so $hf_1 = \phi_1$ (for metal 1)

When $f = f_2$, KE $= 0$ for metal 2,

so $hf_2 = \phi_2$ (for metal 2)

But $f_2 = 3f_1$... So ϕ_2 (for metal 2) $= 3hf_1$

When $f = 4f_1$, $KE_{max} = h(4f_1) - \phi$

So, for metal 1:

$KE_{max} = h(4f_1) - \phi_1 = 4hf_1 - hf_1 = 3hf_1$

and for metal 2:

$KE_{max} = h(4f_1) - \phi_2 = 4hf_1 - 3hf_1 = hf_1$

Therefore, the difference in KE $= 3hf_1 - hf_1 = 2hf_1$

Short-answer solutions

Question 21

a $\tau = Fr$
   $4 = F(1.5)$
   $F = 6\,\text{N}$

- 2 marks: correctly calculates the force needed
- 1 mark: substitutes into a relevant equation

b The period, T, is the same for both Riley and Rex.

Riley's velocity is:

$$v = \frac{2\pi r}{T}$$

$$= \frac{2(3.14)(1)}{T}$$

$$= \frac{6.28}{T}$$

Rex's velocity is:

$$v = \frac{2\pi r}{T}$$

$$= \frac{2(3.14)(1.5)}{T}$$

$$= \frac{9.42}{T}$$

Rex's velocity is greater than Riley's, which supports the statement.

- 2 marks: shows mathematically that the dinosaur has a larger tangential velocity
- 1 mark: calculates the velocity of the dinosaur

Question 22

a Because each square is 5.0 m, the projectile travels 10.0 m in 1.0 s; therefore $u_x = 10\,\mathrm{m\,s^{-1}}$

> - 1 mark: determines the horizontal velocity

b In 2 s, the projectile travels $10 \times 5.0\,\mathrm{m} = 50\,\mathrm{m}$.

$$s = ut + \frac{1}{2}at^2$$

$$50 = u(2) + \frac{1}{2}(-9.8)(2)^2$$

$$u = \frac{50 + (4.9)(4)}{2}$$

$$u = 35\,\mathrm{m\,s^{-1}}$$

> - 2 marks: determines the initial vertical velocity
> - 1 mark: makes a partially correct substitution into the correct equation

c $v^2 = u^2 + 2as$

$$0 = (35)^2 + 2(-9.8)s$$

$$s = 63\,\mathrm{m}$$

> - 2 marks: determines the maximum height
> - 1 mark: makes a partially correct substitution into the correct equation

Question 23

a Polarisation can be explained when light is described as an electromagnetic wave where only one plane of the electric field passes through the polariser.

> - 2 marks: outlines how polarisation supports the wave model of light
> - 1 mark: identifies that light is modelled as a wave

b After the polarising filter, the intensity of the light is half the original intensity, i.e. $I_{max} = 50\,\mathrm{lux}$.

The first filter is parallel to the polarising filter, so the intensity is unaffected.

After Filter 2:

$$I_1 = I_{max}\cos^2\theta$$

$$I_2 = I_1\cos^2 20$$

$$I_2 = I_1 0.833$$

$$I_3 = I_2\cos^2 40$$

$$I_3 = I_2 0.587$$

$$I_3 = I_1 0.518$$

Therefore, the light emerging from the third filter is 0.518 times the light intensity through the first filter of 50 lux.

$$0.518 \times 50 = 25.9\,\mathrm{lux}$$

> - 3 marks: correctly calculates the intensity of light emerging from the filters
> - 2 marks: makes a correct calculation
> - 1 mark: makes a substitution into a relevant formula

Question 24 ©NESA 2016 MARKING GUIDELINES SIB Q25

a Both graphs show that the distance increases, the force decreases. However, in Team A's graph, the force between the masses decreases at a decreasing rate, whereas in Team B's graph, the force decreases at a constant rate.

- 2 marks: provides valid comparison between force and distance in the graphs
- 1 mark: identifies a relationship between force and distance in one of the graphs

b Team A's data set has a good range but too few measurements for a valid relationship to be deduced. Team B's data set has sufficient measurements but over an insufficient range of distances for a valid relationship to be deduced.

- 3 marks: makes an informed judgement of the appropriateness of each data set
- 2 marks: identifies the strengths and/or weaknesses of the data set(s)
- 1 mark: provides some relevant information

Question 25

a
$$\frac{F}{l} = \frac{\mu_0}{2\pi} \frac{I_1 I_2}{d}$$

$$8.0 \times 10^{-5} = \frac{4\pi \times 10^{-7}}{2\pi} \frac{3I^2}{0.05}$$

$$I^2 = \frac{(8.0 \times 10^{-5})(0.05)}{3 \times 2 \times 10^{-7}}$$

$$I = 2.58 \, \text{A}$$

- 2 marks: correctly calculates the current
- 1 mark: makes a substitution into a relevant equation

b $F_{\text{due to 1}} = F_{\text{due to 2}}$ (per unit length)

$$8.0 \times 10^{-5} = \frac{\mu_0 I_2 I_3}{2\pi d}$$

$$8.0 \times 10^{-5} = 2 \times 10^{-7} \frac{I^2}{d}$$

$$8.0 \times 10^{-5} = 2 \times 10^{-7} \frac{2.58^2}{d}$$

$$400 = \frac{2.58^2}{d}$$

$$d = 0.0166 \text{ so } 0.017 \text{ m below the wire}$$

The current must be towards the right.

- 3 marks: correctly calculates the distance of current and that the wire should be placed below wire 1/ above wire 2
- 2 marks: attempts to correctly calculate the distance **or** substitutes into the correct formula and identifies the correct placement of the wire
- 1 mark: provides some relevant information

Question 26

Scientific models are based on observations and must be able to predict new behaviour that can be tested by further experimentation. However, if evidence arises that is inconsistent with a model, the model needs to be revised.

For example, Thomson's plum pudding model suggested a positively charged large sphere with negatively charged electrons embedded in it.

Geiger and Marsden's gold foil experiment, in which some alpha particles were observed to bounce back, was inconsistent with that model. The result was the planetary model of the atom.

Electrons moving in a circular orbit are accelerating, and therefore should emit electromagnetic radiation. This does not occur and shows an inconsistency in Rutherford's model.

The result was Bohr's discrete energy planetary model (quantum model). It had the advantage of being able to explain hydrogen spectral lines. However, it could not explain why electrons had discrete energy. The model was then refined by de Broglie and subsequently Schrödinger, who argued that electrons act as standing waves. This model was validated by the electron diffraction experiment by Davisson and Germer.

- 5 marks: explains the progression of all of the atomic models by stating their problems, observations and how each new model enhanced the previous one
- 4 marks: explains the progression of some of the atomic models by stating their problems, observations and how each new model enhanced the previous one
- 3 marks: describes the progression through observations that led to new models
- 2 marks: describes the models
- 1 mark: provides some relevant information

Question 27 ©NESA 2019 MARKING GUIDELINES SII Q31

a After being switched on, the fan exerts a downward force on the air and due to Newton's 3rd law an equal upward force is exerted on the fan by the air. This reduces the net vertical force observed on the spring balance. This effect increases as the fan's speed increases.

Since the fan increases in speed until reaching its maximum after 10 seconds, the force observed on the spring balance will decrease until it reaches a minimum at 10 seconds, after which it remains constant because the forces are balanced.

- 4 marks: explains the changes observed on the spring balance
- 3 marks: explains a change observed on the spring balance **or** relates changes observed on the spring balance to forces acting on the fan
- 2 marks: identifies changes observed on the spring balance and/or forces acting on the fan
- 1 mark: provides some relevant information

b Between 0–10 s, the student's prediction incorrectly shows an increasing current. During this time the magnitude of back EMF in the motor is increasing, therefore reducing the current in the motor.

From 10–15 s, the student's prediction correctly shows a constant current. Since the fan has reached a constant speed, the magnitude of the back EMF is also constant, so the net current in the motor is constant.

- 4 marks: assesses features of the prediction
- 3 marks: assesses a feature of the prediction **or** outlines issues with the prediction
- 2 marks: outlines an issue with the prediction
- 1 mark: provides some relevant information

Question 28

Galileo's method is best described as a time-of-flight methodology. By measuring the time taken, t, for a light signal to move from one position to another (distance d), the speed of light can be determined using $v = \dfrac{d}{t}$.

The microwave methodology relies instead on the wave equation and the concept of standing waves. By measuring the nodal and antinodal points (where the chocolate melts), the wavelength can be determined. With the frequency already known, the speed can be determined using $v = f\lambda$.

Galileo's method is very imprecise, and required precision time measurements, which the technology of his time could not provide. Thus his results were very inaccurate with a very large uncertainty.

The microwave method, however, is more precise and can give results with a much smaller uncertainty.

- 4 marks: provides comparisons between the two methods
- 3 marks: provides one comparison between the two methods
- 2 marks: identifies one aspect of each method
- 1 mark: identifies one aspect of a method

Question 29

Because of the conservation of energy, you cannot get out more energy than you put in. Although it seems like there is little energy being input to the system and a lot being released, all parts of the system must be considered.

A large atom can be split by firing a neutron that is moving at the correct speed. The resulting atoms have a smaller combined mass than the mass of the original atom and neutron. This means some of the mass has been converted to energy according to $E = mc^2$.

$E = mc^2$ demonstrates the equivalence between mass and energy. Thus, the conservation law can be better stated as conservation of mass–energy.

Therefore, the statement is actually incorrect. Nuclear energy, with the conversion of mass into energy, is consistent with the refined conservation law.

- 3 marks: relates the conservation of energy and mass to the energy output to state that the statement is incorrect
- 2 marks: refers to the loss of mass as the source of energy
- 1 mark: provides some relevant information

Question 30

a
$$\lambda = \frac{\ln 2}{t_{\frac{1}{2}}} = \frac{\ln 2}{1} = 0.6931$$

$$N = N_0 e^{-\lambda t}$$

$$22 = 304 e^{-(0.6931)t}$$

$$\ln\left(\frac{22}{304}\right) = -(0.6931)t$$

$$t = 3.79$$

$$\text{time} = 6441 \text{ years}$$

- 3 marks: correctly calculates the time that has passed
- 2 marks: correctly calculates the number of half-lives that have passed
- 1 mark: substitutes into a relevant formula

b In this model, as the coins turn tails up, they are removed. This models that the number of atoms is decreasing, when in radioactive decay the original atoms would be replaced by daughter atoms so the number of atoms would not decrease.

Alternatively, the number of half-lives can only be simulated in integer amounts so there is no way of determining values in between shakes.

- 2 marks: discusses a limitation of the model
- 1 mark: states limitation of the model

Question 31

a Reading from the graph, the period is 7 days and the radius is 7.5 JD.

- 2 marks: determines the radius and period
- 1 mark: determines the radius or period

b The period of Ganymede's orbit is 7 days.

The radius of Ganymede's orbit is 7.5 JD = 1.05×10^9 m

$$\frac{r^3}{T^2} = \frac{GM}{4\pi^2}$$
$$\frac{(1.05 \times 10^9)^3}{(7 \times 3600 \times 24)^2} = \frac{(6.67 \times 10^{-11})M}{4\pi^2}$$
$$M = 1.87 \times 10^{27}\,\text{kg}$$

- 2 marks: uses Kepler's third law to determine the mass of Jupiter
- 1 mark: makes a partially correct substitution into the correct equation

c The moon's period and orbital radius are linked by $\frac{r^3}{T^2}$. This means that if a moon has a smaller orbital radius, its period will be shorter.

Therefore, a close moon will develop a graph that has a smaller amplitude, and its period will also be smaller.

Students could also answer that the radius is larger and therefore the period is longer.

- 2 marks: describes changes to the graph when the radius is smaller and the period is shorter
- 1 mark: identifies that a smaller radius would result in a smaller period

Question 32 ©NESA 2019 MARKING GUIDELINES SII Q35

The constant deflection of the pendulum to the right indicates that the car has a uniform acceleration to the left, and is therefore travelling in a uniform circular motion.

Larger values of θ indicate smaller radii of motion.

The radius of motion can be expressed in terms of θ.

$$T_y = T\cos\theta = mg \qquad\qquad (1)$$

$$T_x = T\sin\theta = ma = \frac{mv^2}{r} \qquad (2)$$

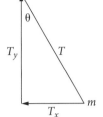

$$\text{from (1)} \quad T = \frac{mg}{\cos\theta}$$

Substituting into (2)

$$\frac{mg\sin\theta}{\cos\theta} = \frac{mv^2}{r}$$

$$\therefore \; g\tan\theta = \frac{v^2}{r}$$

$$r = \frac{v^2}{g\tan\theta}$$

- 4 marks: relates observations to features of the car's motion **and** derives an expression that relates the radius in terms of θ
- 3 marks: relates observations to features of the car's motion **and** provides some steps towards determining an expression for a feature of the motion in terms of θ
- 2 marks: relates an observation to features of the car's motion
- 1 mark: provides some relevant information

Question 33

A black body is an ideal body that absorbs all energy that falls on it and re-emits it as radiation. Using cavities as black body analogues, scientists in the 19th century were able to study the link between the range of wavelengths emitted, the intensity of these wavelengths and the temperature.

When intensity (spectral radiance) versus wavelength is graphed, it produces a characteristic curve called the black body curve.

Wien used this data to establish a relationship between the peak wavelength and temperature, noting an inverse relationship. This allowed astronomers to develop a crude methodology to measure the temperature of stars. However, there was no mathematical model that could explain the shape of the graph, which was based on experimental data.

Planck surmised that black bodies were made of discrete particles and thus should radiate their energy in discrete amounts, according to $E = hf$. He then revised known mathematical models of the time to allow for the energy to be discrete rather than continuous. The result was that the mathematical model matched the experimental data. His solution to the black body curve problem – that energy is discrete – transformed physics.

Einstein used this idea to explain the photoelectric effect, showing that light travelled as particles (later called photons).

Bohr, seeking to resolve a problem in the Rutherford model of the atom, used Planck's idea to establish that electrons exist in discrete orbits and that their angular momentum is quantised.

The work of Planck, Einstein and Bohr ushered in the age of quantum mechanics, which fundamentally changed our understanding of physics.

We continue to use the black body curve in the study of stars and galaxies, and, importantly, the black body curve of the cosmic background radiation allows scientists to not only validate the Big Bang theory but also allows for precise measurements of the temperature of the cosmic background radiation.

- 9 marks: demonstrates an extensive understanding of the black body curve, its origins, the implications, and uses; is clear and concise, using correct scientific terminology
- 7–8 marks: demonstrates a thorough understanding of the black body curve and its origins and implications **and/or** uses
- 5–6 marks: demonstrates a sound understanding of the black body curve and its origins and implications **and/or** uses
- 3–4 marks: outlines what a black body curve is **and/or** identifies some implications/uses
- 2 marks: identifies what a black body curve is
- 1 mark: provides some relevant information

Question 34

a $E = \dfrac{V}{d}$ and $F = Eq$

$F = \dfrac{qV}{d}$

$a = \dfrac{qV}{md}$

$s = \dfrac{1}{2}at^2$

$\dfrac{d}{2} = \dfrac{1}{2}\dfrac{qV}{md}t^2$

$t^2 = \dfrac{md^2}{qV}$

$t = d\sqrt{\dfrac{m}{qV}}$

- 3 marks: derives the expression
- 2 marks: determines the acceleration
- 1 mark: provides a correct substitution into an appropriate equation

b $s = 0\,\text{m s}^{-1}$

$u_y = u\sin 45$

$u_x = u\cos 45$

$a = \dfrac{qV}{md}$

$t = t$

For the maximum height:

$v = u + at$

$0 = -u\sin 45 + \dfrac{qV}{md}t$

$t = u\sin 45\left(\dfrac{md}{qV}\right)$

Since $R = u_x t$, where t represents time up and down,

$R = (u\cos 45)(2)u\sin 45\left(\dfrac{md}{qV}\right)$

$= 2\left(\dfrac{u}{\sqrt{2}}\right)\left(\dfrac{u}{\sqrt{2}}\right)\left(\dfrac{md}{qV}\right)$

$= u^2\left(\dfrac{md}{qV}\right)$

- 3 marks: determines the range
- 2 marks: determines the time
- 1 mark: provides a correct substitution into an appropriate equation

Question 35

The Standard Model was developed in the early 1960s to explain the existence of many subatomic particles that had been detected but that could not be explained by the existing model of the atom involving only protons, neutrons and electrons.

It predicted the existence of fundamental particles – quarks, leptons and bosons – to explain the diversity of subatomic particles that are made up of these fundamental particles.

However, this model, like all scientific models, needed to be validated by experimental evidence that is consistent with the model. The prediction of the existence of the Higgs boson in 1964 needed to be validated, which was done in 2012 with the results from the LHC. Later predictions of its mass (approximately 125 GeV) were also confirmed by the same result from the LHC.

This demonstrates how evidence from particle accelerators can be used to validate the Standard Model.

- 6 marks: uses extensive descriptions of examples to support the use of evidence to understand the Standard Model
- 5 marks: uses extensive description of one example to support the use of evidence to understand the Standard Model
- 4 marks: provides a thorough description with examples to support the use of evidence to understand the Standard Model
- 3 marks: outlines the relevance of evidence in our understanding of the science
- 2 marks: identifies evidence in relation to the development of the Standard Model
- 1 mark: provides some relevant information

HIGHER SCHOOL CERTIFICATE EXAMINATION

Data sheet

Charge on electron, q_e	-1.602×10^{-19} C
Mass of electron, m_e	9.109×10^{-31} kg
Mass of neutron, m_n	1.675×10^{-27} kg
Mass of proton, m_p	1.673×10^{-27} kg
Speed of sound in air	$340 \, \text{m s}^{-1}$
Earth's gravitational acceleration, g	$9.8 \, \text{m s}^{-2}$
Speed of light, c	$3.00 \times 10^8 \, \text{m s}^{-1}$
Electric permittivity constant, ε_0	$8.854 \times 10^{-12} \, \text{A}^2 \text{s}^4 \text{kg}^{-1} \text{m}^{-3}$
Magnetic permeability constant, μ_0	$4\pi \times 10^{-7} \, \text{N A}^{-2}$
Universal gravitational constant, G	$6.67 \times 10^{-11} \, \text{N m}^2 \text{kg}^{-2}$
Mass of Earth, M_E	6.0×10^{24} kg
Radius of Earth, r_E	6.371×10^6 m
Planck constant, h	6.626×10^{-34} J s
Rydberg constant, R (hydrogen)	$1.097 \times 10^7 \, \text{m}^{-1}$
Atomic mass unit, u	1.661×10^{-27} kg
	$931.5 \, \text{MeV}/c^2$
1 eV	1.602×10^{-19} J
Density of water, ρ	$1.00 \times 10^3 \, \text{kg m}^{-3}$
Specific heat capacity of water	$4.18 \times 10^3 \, \text{J kg}^{-1} \text{K}^{-1}$
Wien's displacement constant, b	2.898×10^{-3} m K

Formulae sheet

Motion, forces and gravity

$s = ut + \dfrac{1}{2}at^2$

$v = u + at$

$v^2 = u^2 + 2as$

$\vec{F}_{net} = m\vec{a}$

$\Delta U = mg\Delta h$

$W = F_{\parallel}s = Fs\cos\theta$

$P = \dfrac{\Delta E}{\Delta t}$

$K = \dfrac{1}{2}mv^2$

$\sum \dfrac{1}{2}mv_{before}^2 = \sum \dfrac{1}{2}mv_{after}^2$

$P = F_{\parallel}v = Fv\cos\theta$

$\Delta \vec{p} = \vec{F}_{net}\Delta t$

$\sum \dfrac{1}{2}m\vec{v}_{before} = \sum \dfrac{1}{2}m\vec{v}_{after}$

$\omega = \dfrac{\Delta\theta}{t}$

$a_c = \dfrac{v^2}{r}$

$\tau = r_{\perp}F = rF\sin\theta$

$F_c = \dfrac{mv^2}{r}$

$v = \dfrac{2\pi r}{T}$

$F = \dfrac{GMm}{r^2}$

$U = -\dfrac{GMm}{r}$

$\dfrac{r^3}{T^2} = \dfrac{GM}{4\pi^2}$

Waves and thermodynamics

$v = f\lambda$

$f_{beat} = |\, f_2 - f_1 \,|$

$f = \dfrac{1}{T}$

$f' = f\dfrac{(v_{wave} + v_{observer})}{(v_{wave} - v_{observer})}$

$d\sin\theta = m\lambda$

$n_1\sin\theta_1 = n_2\sin\theta_2$

$n_x = \dfrac{c}{v_x}$

$\sin\theta_c = \dfrac{n_2}{n_1}$

$I = I_{max}\cos^2\theta$

$I_1 r_1^2 = I_2 r_2^2$

$Q = mc\Delta T$

$\dfrac{Q}{t} = \dfrac{kA\Delta T}{d}$

Formulae sheet (continued)

Electricity and magnetism

$E = \dfrac{V}{d}$

$\vec{F} = q\vec{E}$

$V = \dfrac{\Delta U}{q}$

$F = \dfrac{1}{4\pi\varepsilon_0}\dfrac{q_1 q_2}{r^2}$

$W = qV$

$I = \dfrac{q}{t}$

$W = qEd$

$V = IR$

$B = \dfrac{\mu_0 I}{2\pi r}$

$P = VI$

$B = \dfrac{\mu_0 NI}{L}$

$F = qv_\perp B = qvB\sin\theta$

$\Phi = B_\parallel A = BA\cos\theta$

$F = lI_\perp B = lIB\sin\theta$

$\varepsilon = -N\dfrac{\Delta\phi}{\Delta t}$

$\dfrac{F}{l} = \dfrac{\mu_0}{2\pi}\dfrac{I_1 I_2}{r}$

$\dfrac{V_\text{p}}{V_\text{s}} = \dfrac{N_\text{p}}{N_\text{s}}$

$\tau = nIA_\perp B = nIAB\sin\theta$

$V_\text{p}I_\text{p} = V_\text{s}I_\text{s}$

Quantum, special relativity and nuclear

$\lambda = \dfrac{h}{mv}$

$t = \dfrac{t_0}{\sqrt{\left(1 - \dfrac{v^2}{c^2}\right)}}$

$K_{\max} = hf - \phi$

$\lambda_{\max} = \dfrac{b}{T}$

$l = l_0\sqrt{\left(1 - \dfrac{v^2}{c^2}\right)}$

$E = mc^2$

$E = hf$

$p_v = \dfrac{m_0 v}{\sqrt{\left(1 - \dfrac{v^2}{c^2}\right)}}$

$\dfrac{1}{\lambda} = R\left(\dfrac{1}{n_\text{f}^2} - \dfrac{1}{n_\text{i}^2}\right)$

$N_t = N_0 e^{-\lambda t}$

$\lambda = \dfrac{\ln 2}{t_{\frac{1}{2}}}$

9780170465298

Periodic Table of the Elements

KEY

79
Au
197.0
Gold

Atomic Number
Symbol
Standard Atomic Weight
Name

1																	2
H 1.008 Hydrogen																	He 4.003 Helium
3 Li 6.941 Lithium	4 Be 9.012 Beryllium											5 B 10.81 Boron	6 C 12.01 Carbon	7 N 14.01 Nitrogen	8 O 16.00 Oxygen	9 F 19.00 Fluorine	10 Ne 20.18 Neon
11 Na 22.99 Sodium	12 Mg 24.31 Magnesium											13 Al 26.98 Aluminium	14 Si 28.09 Silicon	15 P 30.97 Phosphorus	16 S 32.07 Sulfur	17 Cl 35.45 Chlorine	18 Ar 39.95 Argon
19 K 39.10 Potassium	20 Ca 40.08 Calcium	21 Sc 44.96 Scandium	22 Ti 47.87 Titanium	23 V 50.94 Vanadium	24 Cr 52.00 Chromium	25 Mn 54.94 Manganese	26 Fe 55.85 Iron	27 Co 58.93 Cobalt	28 Ni 58.69 Nickel	29 Cu 63.55 Copper	30 Zn 65.38 Zinc	31 Ga 69.72 Gallium	32 Ge 72.64 Germanium	33 As 74.92 Arsenic	34 Se 78.96 Selenium	35 Br 79.90 Bromine	36 Kr 83.80 Krypton
37 Rb 85.47 Rubidium	38 Sr 87.61 Strontium	39 Y 88.91 Yttrium	40 Zr 91.22 Zirconium	41 Nb 92.91 Niobium	42 Mo 95.96 Molybdenum	43 Tc Technetium	44 Ru 101.1 Ruthenium	45 Rh 102.9 Rhodium	46 Pd 106.4 Palladium	47 Ag 107.9 Silver	48 Cd 112.4 Cadmium	49 In 114.8 Indium	50 Sn 118.7 Tin	51 Sb 121.8 Antimony	52 Te 127.6 Tellurium	53 I 126.9 Iodine	54 Xe 131.3 Xenon
55 Cs 132.9 Caesium	56 Ba 137.3 Barium	57–71 Lanthanoids	72 Hf 178.5 Hafnium	73 Ta 180.9 Tantalum	74 W 183.9 Tungsten	75 Re 186.2 Rhenium	76 Os 190.2 Osmium	77 Ir 192.2 Iridium	78 Pt 195.1 Platinum	79 Au 197.0 Gold	80 Hg 200.6 Mercury	81 Tl 204.4 Thallium	82 Pb 207.2 Lead	83 Bi 209.0 Bismuth	84 Po Polonium	85 At Astatine	86 Rn Radon
87 Fr Francium	88 Ra Radium	89–103 Actinoids	104 Rf Rutherfordium	105 Db Dubnium	106 Sg Seaborgium	107 Bh Bohrium	108 Hs Hassium	109 Mt Meitnerium	110 Ds Darmstadtium	111 Rg Roentgenium	112 Cn Copernicium	113 Nh Nihonium	114 Fl Flerovium	115 Mc Moscovium	116 Lv Livermorium	117 Ts Tennessine	118 Og Oganesson

Lanthanoids

57 La 138.9 Lanthanum	58 Ce 140.1 Cerium	59 Pr 140.9 Praseodymium	60 Nd 144.2 Neodymium	61 Pm Promethium	62 Sm 150.4 Samarium	63 Eu 152.0 Europium	64 Gd 157.3 Gadolinium	65 Tb 158.9 Terbium	66 Dy 162.5 Dysprosium	67 Ho 164.9 Holmium	68 Er 167.3 Erbium	69 Tm 168.9 Thulium	70 Yb 173.1 Ytterbium	71 Lu 175.0 Lutetium

Actinoids

89 Ac Actinium	90 Th 232.0 Thorium	91 Pa 231.0 Protactinium	92 U 238.0 Uranium	93 Np Neptunium	94 Pu Plutonium	95 Am Americium	96 Cm Curium	97 Bk Berkelium	98 Cf Californium	99 Es Einsteinium	100 Fm Fermium	101 Md Mendelevium	102 No Nobelium	103 Lr Lawrencium

Standard atomic weights are abridged to four significant figures.

Elements with no reported values in the tables have no stable nuclides.

Information on elements with atomic numbers 113 and above is sourced from the International Union of Pure and Applied Chemistry Periodic Table of Elements (November 2016 version).

The International Union of Pure and Applied Chemistry Periodic Table of the Elements (February 2010 version) is the principal source of all other data. Some data may have been modified.